Student Solutions Manual and Study Guide

College Physics

Sixth Edition
Volume I

John R. Gordon
James Madison University

Charles Teague
Eastern Kentucky University

Raymond A. Serway
James Madison University

THOMSON

BROOKS/COLE

Australia • Canada • Mexico • Singapore • Spain • United Kingdom • United States

Printed in the United States of America
2 3 4 5 6 7 05 04 03

Printer: Patterson Printing

0-03-034811-0

For more information about our products,
contact us at:
Thomson Learning Academic Resource Center
1-800-423-0563

For permission to use material from this text,
contact us by:
Phone: 1-800-730-2214
Fax: 1-800-731-2215
Web: http://www.thomsonrights.com

Asia
Thomson Learning
5 Shenton Way #01-01
UIC Building
Singapore 068808

Australia
Nelson Thomson Learning
102 Dodds Street
South Street
South Melbourne, Victoria 3205
Australia

Canada
Nelson Thomson Learning
1120 Birchmount Road
Toronto, Ontario M1K 5G4
Canada

Europe/Middle East/South Africa
Thomson Learning
High Holborn House
50/51 Bedford Row
London WC1R 4LR
United Kingdom

Latin America
Thomson Learning
Seneca, 53
Colonia Polanco
11560 Mexico D.F.
Mexico

Spain
Paraninfo Thomson Learning
Calle/Magallanes, 25
28015 Madrid, Spain

Table of Contents

Preface

This <u>Student Solutions Manual and Study Guide</u> has been written to accompany the textbook **College Physics**, Sixth Edition, by Raymond A. Serway and Jerry S. Faughn. The purpose of this Study Guide is to provide the students with a convenient review of the basic concepts and applications presented in the textbook, together with solutions to selected end-of-chapter problems from the textbook. The Study Guide is not an attempt to rewrite the textbook in a condensed fashion. Rather, emphasis is placed upon clarifying typical troublesome points, and providing further drill in methods of problem solving.

Each chapter of the Study Guide is divided into several parts, and every textbook chapter has a matching chapter in the Study Guide. Very often, reference is made to specific equations or figures in the textbook. Every feature of the Study Guide has been included to insure that it serves as a useful supplement to the textbook. Most chapters contain the following sections:

- **Notes From Selected Chapter Sections:** This is a summary of important concepts, newly defined physical quantities, and rules governing their behavior.

- **Equations and Concepts:** This represents a review of the chapter, with emphasis on highlighting important concepts and describing important equations and formalisms.

- **Suggestions, Skills, and Strategies:** This offers hints and strategies for solving typical problems that you will often encounter in the course. In some sections, suggestions are made concerning mathematical skills that are necessary in the analysis of problems.

- **Review Checklist:** This is a list of topics and techniques that you should master after reading the chapter and working the assigned problems.

- **Solutions to Selected End-of-Chapter Problems:** Solutions are shown for approximately twenty percent of the problems from each chapter of the text. Problems were selected to illustrate important concepts in each chapter.

We sincerely hope that this Study Guide will be useful to you in reviewing the material presented in the text, and in improving your ability to solve problems and score well on exams. We welcome any comments or suggestions which could help improve the content of this study guide in future editions; and we wish you success in your study.

John R. Gordon,
Harrisonburg, VA 22807

Charles D. Teague,
Richmond, KY 40475

Raymond A. Serway,
Leesburg, VA 20176

Acknowledgments

It is a pleasure to acknowledge the excellent work of Michael Rudmin, Jonas Šiaulys, and Aleksandras Urbonas of Diversified Service Company—Publishing, whose attention to detail in the preparation of the camera-ready copy did much to enhance the quality of this Sixth Edition of the Student Solutions Manual and Study Guide to accompany College Physics. Their graphics skills and technical expertise combined to produce illustrations for earlier editions which continue to add much to the appearance and usefulness of this volume.

Special thanks go to Senior Developmental Editor, Susan Dust Pashos and Assistant Editor for Chemistry and Physics, Alyssa White of Brooks/Cole Thomson Learning for managing all phases of this project. Randall Jones of Loyola College in Maryland served as accuracy reviewer for this volume and made many helpful suggestions. Finally, we express our appreciation to our families for their inspiration, patience, and encouragement.

Suggestions for Study

Very often we are asked "How should I study this subject, and prepare for examinations?" There is no simple answer to this question; however, we would like to offer some suggestions which may be useful to you.

1. It is essential that you understand the basic concepts and principles before attempting to solve assigned problems. This is best accomplished through a careful reading of the textbook before attending your lecture on that material, jotting down certain points which are not clear to you, taking careful notes in class, and asking questions. You should reduce memorization of material to a minimum. Memorizing sections of a text, equations, and derivations does not necessarily mean you understand the material. Perhaps the best test of your understanding of the material will be your ability to solve the problems in the text, or those given on exams.

2. Try to solve as many problems at the end of the chapter as possible. You will be able to check the accuracy of your calculations to the odd-numbered problems, since the answers to these are given at the back of the text. Furthermore, detailed solutions to approximately twenty percent of the problems from the text are provided in this Study Guide. Many of the worked examples in the text will serve as a basis for your study.

3. The method of solving problems should be carefully planned. First, read the problem several times until you are confident you understand what is being asked. Look for key words which will help simplify the problem, and perhaps allow you to make certain assumptions. You should also pay special attention to the information provided in the problem.

    It is a good idea to write down the given information before proceeding with a solution. (For example, $a = -3.00$ m/s^2 and $v_0 = 5.00$ m/s are given. Find the velocity v, and the displacement Δx after $\Delta t = 2.00$ s.) After you have decided on the method you feel is appropriate for the problem, proceed with your solution. If you are having difficulty in working problems, we suggest that you again read the text and your lecture notes. It may take several readings before you are ready to solve certain problems. The solved problems in this Study Guide should be of value to you in this regard.

4. After reading a chapter, you should be able to define any new quantities that were introduced, and discuss the first principles that were used to derive fundamental formulas. A review is provided in each chapter of the Study Guide for this purpose, and the marginal notes in the textbook (or the index) will help you locate these topics. You should be able to correctly associate with each physical quantity the symbol used to represent that quantity (including vector notation, if appropriate) and the SI unit in which the quantity is specified. Furthermore, you should be able to express each important formula or equation in a concise and accurate prose statement.

5. We suggest that you use this Study Guide to review the material covered in the text, and as a guide in preparing for exams. You should also use the **Chapter Review, Notes From Selected Chapter Sections,** and **Equations and Concepts** to focus in on any points which require further study. Remember that the main purpose of this Study Guide is to improve upon the efficiency and effectiveness of your study hours and your overall understanding of physical concepts. However, it should not be regarded as a substitute for your textbook or individual study and practice in problem solving.

6. The user of this study guide should keep in mind the procedure that has been followed when displaying intermediate results during a problem solution. When, for the sake of clarity, it was desirable to display an intermediate result, that result has been rounded off to the appropriate number of significant figures. *However, many additional digits of that result were retained in the memory of the calculator and used in subsequent calculations.* If this procedure was not followed, the final answer obtained could vary, dependent on how many intermediate results were displayed (with associated round off error in each) and the stages in the solution where it was decided to provide intermediate results.

Chapter 1

INTRODUCTION

NOTES ON SELECTED CHAPTER SECTIONS

1.1   Standards of Length, Mass, and Time

Systems of units commonly used are the **SI system**, in which the units of mass, length, and time are the kilogram (kg), meter (m), and second (s), respectively; the **cgs** or **gaussian system**, in which the units of mass, length, and time are the gram (g), centimeter (cm), and second, respectively; and the U.S. customary system, in which the units of mass, length, and time are the slug, foot (ft), and second, respectively.

In 1960, the meter was defined as 1 650 763.73 wavelengths of orange-red light emitted from a krypton-86 lamp. However, in October 1983, the **meter** was redefined to be **the distance traveled by light in a vacuum during a time of 1/299 792 458 second.**

The SI unit of mass, the **kilogram**, is defined as **the mass of a specific platinum-iridium alloy cylinder kept at the International Bureau of Weights and Measures at Sèvres, France.**

The **second** is now defined as **9 192 631 770 times the period of one oscillation of radiation from the Cesium-133 atom.**

1.2   The Building Blocks of Matter

It is useful to view the atom as a miniature Solar System with a dense, positively charged nucleus occupying the position of the Sun and negatively charged electrons orbiting like the planets. Occupying the nucleus are two basic entities, protons and neutrons. The **proton** is nature's fundamental carrier of positive charge; the **neutron** has no charge and a mass about equal to that of a proton. We shall find in Chapter 30 that even more elementary building blocks than protons and neutrons exist. Protons and neutrons are each now thought to consist of three particles called **quarks**.

1.3   Dimensional Analysis

Dimensional analysis makes use of the fact that **dimensions can be treated as algebraic quantities.** That is, quantities can be added or subtracted only if they have the same dimensions. Furthermore, the **terms** on each side of an equation must have the same dimensions.

1

1.4   Uncertainty in Measurement and Significant Figures

When multiplying several quantities, the number of significant figures in the final **result** is the same as the number of significant figures in the **least** accurate of the quantities being multiplied, where "least accurate" means "having the lowest number of significant figures". The same rule applies to division. When numbers are added (or subtracted), the number of decimal places in the result should equal the smallest number of decimal places of any term in the sum (or difference). Most of the numerical examples and end-of-chapter problems will yield answers having either two or three significant figures.

1.5   Conversion of Units

Sometimes it is necessary to convert units from one system to another. A list of conversion factors can be found on the inside of the back cover of the **Student Solution Manual**.

1.6   Order-of-Magnitude Calculations

Often it is useful to estimate an answer to a problem in which little information is given. In such a case we refer to the **order of magnitude** of a quantity, by which we mean the power of ten that is closest to the actual value of the quantity. Usually, when an order-of-magnitude calculation is made, the results are reliable to within a factor of 10.

1.7   Coordinate Systems

A coordinate system used to specify locations in space consists of:

1. A fixed reference point O, called the origin
2. A set of specified axes, or directions with an appropriate scale and label on each of the axes
3. Instructions that tell us how to label a point in space relative to the origin and axes

1.8   Trigonometry

The portion of mathematics that is based on the special properties of a right triangle is called trigonometry. You should review the basic trigonometric functions stated by Equations 1.1 and 1.2 in the following **Equations and Concepts** section.

1.9   Problem-Solving Strategy

In developing problem-solving strategies, seven basic steps are commonly used:

1. Read the problem carefully at least twice. Be sure you understand the nature of the problem before proceeding further.

2. Draw a suitable diagram with appropriate labels and coordinate axes, if needed.

3. Imagine a movie, running in your mind, of what happens in the problem.

4. As you examine what is being asked in the problem, identify the basic physical principle (or principles) that are involved, listing the knowns and unknowns.

5. Select a basic relationship or derive an equation that can be used to find the unknown, and symbolically solve the equation for the unknown.

6. Substitute the given values with the appropriate units into the equation.

7. Obtain a numerical value with units for the unknown. The problem is verified and receives a check mark if the following questions can be properly answered: Do the units match? Is the answer reasonable? Is the plus or minus sign proper or meaningful?

EQUATIONS AND CONCEPTS

The three most basic trigonometric functions of one of the acute angles of a right triangle are the sine, cosine, and tangent.

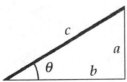

$$\sin \theta = \frac{\text{side opposite to } \theta}{\text{hypotenuse}} = \frac{a}{c}$$

$$\cos \theta = \frac{\text{side adjacent to } \theta}{\text{hypotenuse}} = \frac{b}{c} \qquad (1.1)$$

$$\tan \theta = \frac{\text{side opposite to } \theta}{\text{side adjacent to } \theta} = \frac{a}{b}$$

The Pythagorean theorem is an important relationship among the lengths of the sides of a right triangle. In this equation, c represents the hypotenuse.

$$c^2 = a^2 + b^2 \qquad (1.2)$$

SUGGESTIONS, SKILLS, AND STRATEGIES

Many mathematical symbols will be used throughout this book. Some important examples are listed below.

$\propto$	denotes	a proportionality
$<$	means	"is less than"
$>$	means	"is greater than"
$<<$	means	"is much less than"
$>>$	means	"is much greater than"
$\cong$	indicates	approximate equality
$=$	indicates	equality
Δx ("delta x")	indicates	the change in a quantity x
$\lvert x \rvert$	indicates	the absolute value of x (always positive)
Σ (capital sigma)	represents	a sum. For example,

$$x_1 + x_2 + x_3 + x_4 + x_5 = \sum_{i=1}^{5} x_i$$

REVIEW CHECKLIST

▷ Discuss the units of length, mass and time and the standards for these quantities in SI units. (Section 1.1).

▷ Derive the quantities force, velocity, volume, acceleration, etc. from the three basic quantities.

▷ Perform a dimensional analysis of an equation containing physical quantities whose individual units are known. (Section 1.3)

▷ Convert units from one system to another. (Section 1.5)

▷ Carry out order-of-magnitude calculations or guesstimates. (Section 1.6)

▷ Describe the coordinates of a point in space using a Cartesian coordinate system. (Section 1.8)

SOLUTIONS TO SELECTED END-OF-CHAPTER PROBLEMS

1. In a desperate attempt to come up with an equation to use during an examination, a student tries $v^2 = ax$. Use dimensional analysis to determine whether this equation might be valid.

Solution	v	has units of	(L / T)
	a	has units of	$\left(L / T^2\right)$
and	x	has units of	(L)
Thus, the left side of the equation	v^2	has units of	$(L / T)^2 = L^2 / T^2$
and the right side	ax	has units of	$\left(L / T^2\right)L - L^2 / T^2$

Since the two sides of the equation have the same units, we must conclude that from the standpoint of units alone, the equation might be valid. ◊

---

9. Carry out the following arithmetic operations: (a) the sum of the numbers 756, 37.2, 0.83, and 2.5; (b) the product 0.0032×356.3; (c) the product $5.620 \times \pi$.

Solution

(a) Using a calculator gives a total of 796.53.

In deciding how many significant figures to report in this answer, recall the rule for addition and subtraction:

"When numbers are added or subtracted, the number of decimal places in the result should equal the smallest number of decimal places of any term in the sum."

Thus, the result should be rounded to 797. ◊

(b)   The calculator gives $0.0032 \times 356.3 = \left(3.2 \times 10^{-3}\right) \times 356.3 = 1.14016$

The rule for significant figures in products states that

> "When multiplying two or more quantities, the number of significant figures in the final product is the same as the number of significant figures in the least accurate of the quantities multiplied."

Therefore, the product must be rounded to 1.1 because 3.2×10^{-3} has only two significant figures.   ◊

(c)   The number π is known to a very large number of significant figures, but the product $5.620 \times \pi$ must be rounded to 17.66 because 5.620 has only four significant figures.   ◊

---

21.   The speed of light is about 3.00×10^8 m / s.   Convert this to miles per hour.

Solution

A systematic way to convert units is to multiply the original value by one or more ratios, each ratio chosen so that its value is unity (i.e., its numerator and denominator are equal). Further, each ratio should be chosen so that multiplication by it will cancel some of the current units and move you one step closer to the desired units.

Applying this technique to the stated problem gives:

$$3.00 \times 10^8 \text{ m / s} = \left(3.00 \times 10^8 \text{ m / s}\right)\left(\frac{1 \text{ km}}{10^3 \text{m}}\right)\left(\frac{0.621 \text{ mi}}{1 \text{ km}}\right)\left(\frac{3600 \text{ s}}{1 \text{ h}}\right) = 6.71 \times 10^8 \text{ mi / h} \qquad ◊$$

---

25. A quart container of ice cream is to be made in the form of a cube. What should be the length of a side, in centimeters? (Use the conversion 1 gallon = 3.786 liter.)

Solution

Since the container is in the form of a cube, with sides of length L, its volume may be written as $V = L \times L \times L = L^3 = 1.00$ quart.

Converting to units of cubic centimeters yields:

$$L^3 = (1.00 \text{ quart}) \left(\frac{1 \text{ gallon}}{4 \text{ quart}} \right) \left(\frac{3.786 \text{ liter}}{1 \text{ gallon}} \right) \left(\frac{1000 \text{ cm}^3}{1 \text{ liter}} \right) = 946 \text{ cm}^3$$

Solving for the length of one side then gives

$$L = \sqrt[3]{946 \text{ cm}^3} = 9.82 \text{ cm}$$

◊

31. An automobile tire is rated to last for 50 000 miles. Estimate the number of revolutions the tire will make in its lifetime.

Solution

Tires for full-size automobiles often have radii of 15 or 16 inches, or approximately 1.25 ft.

Thus, the circumference of the tire is $C = 2\pi r \cong 8$ ft, and a reasonable guess for the number of revolutions would be

$$n = \frac{\text{distance traveled}}{\text{circumference}} = \frac{x}{2\pi r} \cong \frac{50\,000 \text{ mi}}{8 \text{ ft / rev}} (5280 \text{ ft / mi}) = 3 \times 10^7 \text{ rev}$$

or $\qquad n \sim 10^7 \text{ rev}$ ◊

35. A point is located in a polar coordinate system by the coordinates $r = 2.5$ m and $\theta = 35°$. Find the x and y coordinates of this point, assuming the two coordinate systems have the same origin.

Solution   The point P is shown in the sketch below with both its polar and Cartesian coordinates labeled. The shaded triangle and basic trigonometry may be used to convert from the polar coordinates to the corresponding Cartesian coordinates:

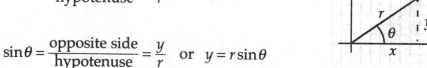

$$\cos\theta = \frac{\text{adjacent side}}{\text{hypotenuse}} = \frac{x}{r} \quad \text{or} \quad x = r\cos\theta$$

$$\sin\theta = \frac{\text{opposite side}}{\text{hypotenuse}} = \frac{y}{r} \quad \text{or} \quad y = r\sin\theta$$

With $r = 2.5$ m   and $\theta = 35°$

these yield: $x = 2.0$ m   and $y = 1.4$ m   ◊

39.   For the triangle shown in Figure P1.39, what are (a) the length of the unknown side, (b) the tangent of θ, and (c) the sin of ϕ?

Solution

(a)   The unknown side, labeled x in the figure, is most easily found using the Pythagorean Theorem,

Figure P1.39

$$(9.00 \text{ m})^2 = x^2 + (6.00 \text{ m})^2$$

giving $x^2 = 45.0$ m^2 or $x = 6.71$ m   ◊

(b) $\tan\theta = \dfrac{\text{side opposite } \theta}{\text{side adjacent to } \theta} = \dfrac{6.00 \text{ m}}{x} = \dfrac{6.00 \text{ m}}{6.71 \text{ m}} = 0.894$ ◊

(c) $\sin\phi = \dfrac{\text{side opposite } \phi}{\text{hypotenuse}} = \dfrac{x}{9.00 \text{ m}} = \dfrac{6.71 \text{ m}}{9.00 \text{ m}} = 0.746$ ◊

42. A right triangle has a hypotenuse of length 3.00 m, and one of its angles is 30.0°. What are the lengths of (a) the side opposite the 30.0° angle and (b) the side adjacent to the 30.0° angle?

Solution

The described triangle is shown in the sketch with the unknown sides labeled a and b. These sides may be found using basic trigonometry.

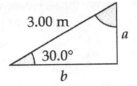

(a) The sine function is chosen here since it relates the unknown side a and the known hypotenuse.

$$\sin 30.0° = \frac{\text{opposite side}}{\text{hypotenuse}} = \frac{a}{3.00 \text{ m}} \quad \text{or} \quad a = (3.00 \text{ m})\sin 30.0° = 1.50 \text{ m} \quad \lozenge$$

(b) Here, the cosine function is most convenient since it will relate the unknown side b to the known hypotenuse.

$$\cos 30.0° = \frac{\text{adjacent side}}{\text{hypotenuse}} = \frac{b}{3.00 \text{ m}} \quad \text{or} \quad b = (3.00 \text{ m})\cos 30.0° = 2.60 \text{ m} \quad \lozenge$$

---

48. The radius of the planet Saturn is 5.85×10^7 m, and its mass is 5.68×10^{26} kg (Fig. P1.48). (a) Find the density of Saturn (its mass divided by its volume) in grams per cubic centimeter. (The volume of a sphere is given by $\frac{4}{3}\pi r^3$.) (b) Find the area of Saturn in square feet. (The surface area of a sphere is given by $4\pi r^2$.)

Solution

(a) The volume of Saturn is $V = \frac{4}{3}\pi r^3 = \frac{4}{3}\pi(5.85 \times 10^7 \text{ m})^3 = 8.39 \times 10^{23} \text{ m}^3$

and the density is
$$\frac{m}{V} = \left(\frac{5.68 \times 10^{26} \text{ kg}}{8.39 \times 10^{23} \text{ m}^3}\right)\left(\frac{1 \text{ m}}{10^2 \text{ cm}}\right)^3\left(\frac{10^3 \text{ g}}{1 \text{ kg}}\right) = 0.677 \text{ g} / \text{cm}^3 \quad \lozenge$$

(b) The surface area of Saturn is

$$A = 4\pi r^2 = 4\pi(5.85 \times 10^7 \text{ m})^2\left(\frac{3.281 \text{ ft}}{1 \text{ m}}\right)^2 = 4.63 \times 10^{17} \text{ ft}^2 \quad \lozenge$$

---

51. You can obtain a rough estimate of the size of a molecule by the following simple experiment. Let a droplet of oil spread out on a smooth water surface. The resulting oil slick will be approximately one molecule thick. Given an oil droplet of mass 9.00×10^{-7} kg and density 918 kg / m^3 that spreads out into a circle of radius 41.8 cm on the water surface, what is the order of magnitude of the diameter of an oil molecule?

Solution

The density of an object is its mass divided by its volume. Recognizing this, the volume of oil in the droplet may be found from the equation

$$\text{density} = \frac{\text{mass}}{\text{volume}} : \qquad \text{volume} = \frac{\text{mass}}{\text{density}} = \frac{9.00 \times 10^{-7} \text{ kg}}{918 \text{ kg / m}^3} = 9.80 \times 10^{-10} \text{ m}^3$$

The oil slick has a cylindrical shape with a height, h, equal to the diameter of an oil molecule, and a circular cross-section of radius $r = 41.8 \text{ cm} = 0.418 \text{ m}$. Its volume is then given by

$$\text{volume} = (\text{height})(\text{cross-sectional area}) :$$

or $$\text{volume} = h\left(\pi r^2\right) = h\pi(0.418 \text{ m})^2 = h\left(0.549 \text{ m}^2\right)$$

Since the volume of oil in the slick is same as the volume of the droplet,

or $$h\left(0.549 \text{ m}^2\right) = 9.80 \times 10^{-10} \text{ m}^3$$

or $$h = \frac{9.80 \times 10^{-10} \text{ m}^3}{0.549 \text{ m}^2} = 1.78 \times 10^{-9} \text{ m}$$

The order of magnitude of the diameter of an oil molecule is then: $\quad h \sim 10^{-9} \text{ m} \qquad \Diamond$

Chapter 2
MOTION IN ONE DIMENSION

NOTES ON SELECTED CHAPTER SECTIONS

2.1 Displacement

The displacement of an object, defined as its **change in position**, is given by the difference between its final and initial coordinates, or $x_f - x_i$. Displacement is an example of a vector quantity. A vector is a physical quantity that requires a specification of both direction and magnitude.

2.2 Average Velocity

The **average velocity** of an object during the time interval t_i to t_f is equal to the slope of the straight line joining the initial and final points on a graph of the position of the object plotted versus time.

2.3 Instantaneous Velocity

The slope of the line tangent to the position-time curve at a point P is defined to be the **instantaneous velocity** at the corresponding time.

The **instantaneous speed** of an object, which is a scalar quantity, is defined as the magnitude of the instantaneous velocity. Hence, by definition, **speed can never be negative**.

2.4 Acceleration

The **average acceleration** during a given time interval is defined as the change in velocity divided by the time interval during which this change occurs.

The **instantaneous acceleration** of an object at a certain time equals the slope of the tangent to the velocity-time graph at that instant of time.

2.5 Motion Diagrams

It is often instructive to make use of motion diagrams to describe the velocity and acceleration vectors as time progresses while an object is in motion. You should carefully study Figure 2.12 in the textbook. This figure illustrates the motion of a car in three different cases:

1. Constant positive velocity, zero acceleration.
2. Positive velocity, positive acceleration.
3. Positive velocity, negative acceleration.

2.6   One-Dimensional Motion with Constant Acceleration

This type of motion is important because it applies to many objects in nature. When an object moves with constant acceleration, the average acceleration equals the instantaneous acceleration. Equations 2.6 through 2.10 may be used to solve any problem in one-dimensional motion with <u>constant acceleration</u>.

2.7   Freely Falling Objects

A freely falling body is an object **moving freely under the influence of gravity only**, regardless of its initial motion.

It is important to emphasize that any freely falling object experiences an **acceleration directed downward**. This is true regardless of the direction of motion of the object.

An object thrown upward (or downward) will experience the same acceleration as an object released from rest. Once they are in free fall, all objects have an acceleration downward equal to the acceleration due to gravity.

EQUATIONS AND CONCEPTS

The average velocity of an object during a time interval is the ratio of the total displacement to the time interval during which the displacement occurred.

$$\bar{v} = \frac{\Delta x}{\Delta t} = \frac{x_f - x_i}{t_f - t_i} \tag{2.2}$$

The instantaneous velocity v is defined as the limit of the average velocity as the time interval Δt becomes infinitesimally short. Note that the instantaneous velocity of an object might have different values from instant to instant.

$$v \equiv \lim_{\Delta t \to 0} \frac{\Delta x}{\Delta t} \tag{2.3}$$

The average acceleration of an object during a time interval is the ratio of the change in velocity to the time interval during which the change in velocity occurs.

$$\bar{a} \equiv \frac{\Delta v}{\Delta t} = \frac{v_f - v_i}{t_f - t_i} \qquad (2.4)$$

The instantaneous acceleration is defined as the limit of the average acceleration as the time interval Δt goes to zero.

$$a \equiv \lim_{\Delta t \to 0} \frac{\Delta v}{\Delta t} \qquad (2.5)$$

These equations are called the equations of kinematics, and can be used to describe one-dimensional motion along the x axis with constant acceleration. Note that each equation shows a different relationship among physical quantities: initial velocity, final velocity, acceleration, time, and displacement. Also in the form that they are shown, it is assumed that $t_i = 0$, $t_f = t$, and for convenience, $\Delta x = x - x_0$.

$$v - v_0 + at \qquad (2.6)$$

$$\bar{v} = \frac{v_0 + v}{2} \qquad (2.7)$$

$$\Delta x = \frac{1}{2}(v + v_0)t \qquad (2.8)$$

$$\Delta x = v_0 t + \frac{1}{2}at^2 \qquad (2.9)$$

$$v^2 = v_0^2 + 2a\Delta x \qquad (2.10)$$

The equations which describe the motion of an object in free fall $(a = -g)$ along the y axis are stated here. Note that since $a = -g$ in these four equations, the $+y$ axis is predefined to point upwards.

$$v = v_0 - gt$$

$$\Delta y = \frac{1}{2}(v + v_0)t$$

$$\Delta y = v_0 t - \frac{1}{2}gt^2$$

$$v^2 = v_0^2 - 2g\Delta y$$

SUGGESTIONS, SKILLS, AND STRATEGIES

The following procedure is recommended for solving problems involving motion with constant acceleration.

1. Make sure all the units in the problem are consistent. That is, if distances are measured in meters, be sure that velocities have units of m/s and accelerations have units of m/s^2.

2. Choose a coordinate system, and make a labeled diagram of the problem including the directions of all displacements, velocities, and accelerations.

3. Make a list of all the quantities given in the problem and a separate list of those to be determined.

4. Select from the list of equations which describe constant acceleration those that will enable you to determine the unknowns.

5. Construct an appropriate motion diagram and check to see if your answers are consistent with the diagram of the problem.

REVIEW CHECKLIST

▷ Define the displacement and average velocity of a particle in motion. (Sections 2.1 and 2.2)

▷ Define the instantaneous velocity and understand how this quantity differs from average velocity. (Section 2.3)

▷ Define average acceleration and instantaneous acceleration. (Section 2.4)

▷ Construct a graph of displacement versus time (given a function such as $x = 5 + 3t - 2t^2$) for a particle in motion along a straight line. From this graph, you should be able to determine both average and instantaneous values of velocity by calculating the slope of the tangent to the graph. (Section 2.6)

▷ Describe what is meant by a body in **free fall** (one moving under the influence of gravity--where air resistance is neglected). Recognize that the equations of constant accelerated motion apply directly to a freely falling object and that the acceleration is then given by $a = -g$ (where $g = 9.8$ m/s^2). (Section 2.7)

▷ Apply the motion equations of this chapter (Equations 2.6 - 2.10) to any situation where the motion occurs under constant acceleration.

SOLUTIONS TO SELECTED END-OF-CHAPTER PROBLEMS

1.   A person travels by car from one city to another with different constant speeds between pairs of cities. She drives for 30.0 min at 80.0 km/h, 12.0 min at 100 km/h, and 45.0 min at 40.0 km/h, and spends 15.0 min eating lunch and buying gas. (a) Determine the average speed for the trip. (b) Determine the distance between the initial and final cities along this route.

Solution

The distance traveled in each of the four distinct time intervals are:

Interval	Duration	Distance traveled
1	$\Delta t_1 = 30.0$ min $= 0.500$ h	$\Delta x_1 = v_1(\Delta t_1) = (80.0 \text{ km / h})(0.500 \text{ h}) = 40.0$ km
2	$\Delta t_2 = 12.0$ min $= 0.200$ h	$\Delta x_2 = v_2(\Delta t_2) = (100 \text{ km / h})(0.200 \text{ h}) = 20.0$ km
3	$\Delta t_3 = 45.0$ min $= 0.750$ h	$\Delta x_3 = v_3(\Delta t_3) = (40 \text{ km / h})(0.750 \text{ h}) = 30.0$ km
4	$\Delta t_4 = 15.0$ min $= 0.250$ h	$\Delta x_4 = v_4(\Delta t_4) = (0)(0.250 \text{ h}) = 0$

Solving part (b) of this problem first,

(b) $\Delta x_{total} = \sum_{i=1}^{4}(\Delta x_i) = (40.0 + 20.0 + 30.0 + 0) \text{ km} = 90.0 \text{ km}$

(a)   The total time for the trip is

$$\Delta t_{total} = \sum_{i=1}^{4}(\Delta t_i) = (0.500 + 0.200 + 0.750 + 0.250) \text{ h} = 1.70 \text{ h}$$

Using $\Delta x_{total} = \bar{v}(\Delta t_{total})$, the average speed for the trip is found to be

$$\bar{v} = \frac{\Delta x_{total}}{\Delta t_{total}} = \frac{90.0 \text{ km}}{1.70 \text{ h}} = 52.9 \text{ km / h} \qquad \Diamond$$

13. In order to qualify for the finals in a racing event, a race car must achieve an average speed of 250 km/h on a track with a total length of 1600 m. If a particular car covers the first half of the track at an average speed of 230 km/h, what minimum average speed must it have in the second half of the event in order to qualify?

Solution

The time required to travel a distance of 1600 m at an average speed of 250 km/h is

$$t_{total} = \frac{x_{total}}{\overline{v}_{total}} = \frac{1600 \text{ m}}{250 \text{ km / h}}\left(\frac{3600 \text{ s / h}}{1000 \text{ m / km}}\right) = 23.0 \text{ s}$$

This is the maximum total elapsed time if the car is to qualify for the race. If the car travels the first 800 m at an average speed of 230 km/h, the time used for the first half of the trip is

$$t_1 = \frac{x_1}{\overline{v}_1} = \frac{800 \text{ m}}{230 \text{ km / h}}\left(\frac{3600 \text{ s / h}}{1000 \text{ m / km}}\right) = 12.5 \text{ s}$$

Thus, if the car is to qualify, the maximum time that can be used on the second half of the trip is: $t_2 = 23.0 \text{ s} - 12.5 \text{ s} = 10.5 \text{ s}$. The average speed required to cover the remaining 800 m in 10.5 s is given by

$$\overline{v}_2 = \frac{x_2}{t_2} = \frac{800 \text{ m}}{10.5 \text{ s}} = 76.2 \text{ m / s}$$

$$\overline{v}_2 = (76.2 \text{ m / s})\left(\frac{1.00 \text{ km / h}}{0.278 \text{ m / s}}\right) = 274 \text{ km / h} \qquad\qquad \Diamond$$

---

17. Find the instantaneous velocities of the tennis player of Figure P2.7 at (a) 0.50 s, (b) 2.0 s, (c) 3.0 s, (d) 4.5 s.

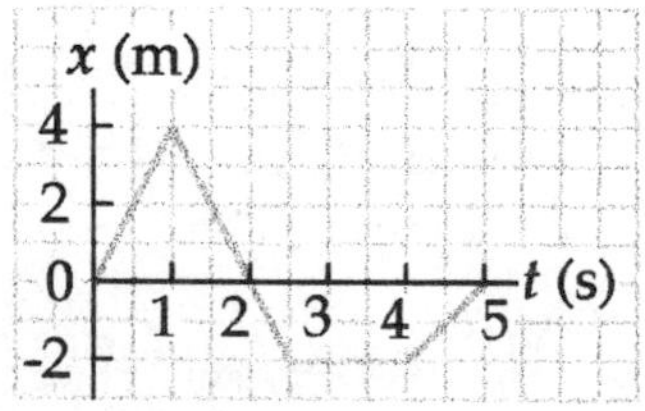

Solution

From Section 2.3 of the textbook, **"The slope of the line tangent to the position-time curve at P is defined to be the instantaneous velocity at that time."**

Figure P2.7

Thus, we need to determine the slope of the tangent lines to the position-time curve shown in Figure P2.7 at each of the requested times.

These slopes may be calculated as follows:

(a) $\quad v_{0.5\,s} = \dfrac{x_{1.0\,s} - x_{0.0\,s}}{1.0\,s - 0.0\,s} = \dfrac{4.0\,m - 0.0\,m}{1.0\,s} = 4.0\,m\,/\,s$ ◊

(b) $\quad v_{2.0\,s} = \dfrac{x_{2.5\,s} - x_{1.0\,s}}{2.5\,s - 1.0\,s} = \dfrac{-2.0\,m - 4.0\,m}{1.5\,s} = \dfrac{-6.0\,m}{1.5\,s} = -4.0\,m\,/\,s$ ◊

(c) $\quad v_{3.0\,s} = \dfrac{x_{4.0\,s} - x_{2.5\,s}}{4.0\,s - 2.5\,s} = \dfrac{-2.0\,m - (-2.0\,m)}{1.5\,s} = \dfrac{0.0\,m}{1.5\,s} = 0.0\,m\,/\,s$ ◊

(d) $\quad v_{4.5\,s} = \dfrac{x_{5.0\,s} - x_{4.0\,s}}{5.0\,s - 4.0\,s} = \dfrac{0.0\,m - (-2.0\,m)}{1.0\,s} = \dfrac{+2.0\,m}{1.0\,s} = 2.0\,m\,/\,s$ ◊

---

21. A certain car is capable of accelerating at a rate of +0.60 m/s². How long does it take for this car to go from a speed of 55 mi/h to a speed of 60 mi/h?

Solution The average acceleration over a time interval of duration Δt is defined as

$$\bar{a} = \frac{\Delta v}{\Delta t}$$

Thus, the time required to achieve a change in velocity Δv, with an average acceleration $\bar{a}$ is

$$\Delta t = \frac{\Delta v}{\bar{a}} = \frac{v_f - v_i}{\bar{a}}$$

Since the car maintains a constant acceleration during the time interval of interest, the average acceleration is the same as the constant instantaneous acceleration (i.e., $\bar{a} = a = +0.60\,m\,/\,s^2$).

The required time is then:

$$\Delta t = \frac{(60\,mi\,/\,h - 55\,mi\,/\,h)}{0.60\,m\,/\,s^2} = \left(\frac{5.0\,mi\,/\,h}{0.60\,m\,/\,s^2}\right)\left(\frac{1609\,m\,/\,mi}{3600\,s\,/\,h}\right) = 3.7\,s \qquad ◊$$

---

29. A Cessna aircraft has a lift-off speed of 120 km/h. (a) What minimum constant acceleration does this require if the aircraft is to be airborne after a takeoff run of 240 m? (b) How long does it take the aircraft to become airborne?

Solution   Assume the aircraft starts from rest ($v_i = 0$) as it accelerates down the runway.

(a) Since the acceleration is uniform, we may use $v_f^2 = v_i^2 + 2a(\Delta x)$ during the take-off run to find

$$a = \frac{v_f^2 - v_i^2}{2\Delta x} = \frac{\left[(120 \text{ km}/\text{h})^2 - 0\right]}{2(240 \text{ m})}\left(\frac{0.278 \text{ m}/\text{s}}{1 \text{ km}/\text{h}}\right)^2 = 2.32 \text{ m}/\text{s}^2 \qquad \lozenge$$

(b) The time required for the aircraft to reach lift-off speed is given by $v_f = v_i + at$ as

$$t = \frac{v_f - v_i}{a} = \left(\frac{120 \text{ km}/\text{h} - 0}{2.32 \text{ m}/\text{s}^2}\right)\left(\frac{0.278 \text{ m}/\text{s}}{1 \text{ km}/\text{h}}\right) = 14.4 \text{ s} \qquad \lozenge$$

35. A train is traveling down a straight track at 20 m/s when the engineer applies the brakes, resulting in an acceleration of $-1.0 \text{ m}/\text{s}^2$ as long as the train is in motion. How far does the train move during a 40-s time interval starting at the instant the brakes are applied?

Solution

It is tempting to try solving this problem by applying the uniformly accelerated motion equation $\Delta x = v_i t + \frac{1}{2}at^2$ to the full 40-s time interval.

This gives $\Delta x = (20 \text{ m}/\text{s})(40 \text{ s}) + \frac{1}{2}\left(-1.0 \text{ m}/\text{s}^2\right)(40 \text{ s})^2 = 0$

which is an obviously incorrect result.

The source of our error may be found by using $v_f = v_i + at$ to find the time required for the train to come to rest.

This yields $t = \dfrac{v_f - v_i}{a} = \dfrac{0 - 20 \text{ m}/\text{s}}{-1.0 \text{ m}/\text{s}^2} = 20 \text{ s}$

Therefore, we see that the train will not have a constant acceleration for the full 40-s time interval, so application of uniformly accelerated motion equations to that time interval is invalid. However, the train does have a constant acceleration during the 20-s interval required for it to come to a stop.

Application of $\Delta x = v_i t + \frac{1}{2}at^2$ to this interval gives the distance traveled while stopping as

$$\Delta x = (20 \text{ m / s})(20 \text{ s}) + \frac{1}{2}\left(-1.0 \text{ m / s}^2\right)(20 \text{ s})^2 = 200 \text{ m} \qquad \Diamond$$

38. A train 400 m long is moving on a straight track with a speed of 82.4 km/h. The engineer applies the brakes at a crossing, and later the last car passes the crossing with a speed of 16.4 km/h. Assuming constant acceleration, determine how long the train blocked the crossing. Disregard the width of the crossing.

Solution

If we ignore the width of the crossing, the displacement of the train from the instant the engine enters the crossing until the last car leaves the crossing is the same as the length of the train;

$\Delta x = 400$ m.

Since the train has a constant acceleration, its average velocity while it is blocking the crossing may be found from:

$$\overline{v} = \frac{v_i + v_f}{2} = \frac{82.4 \text{ km / h} + 16.4 \text{ km / h}}{2} = 49.4 \text{ km / h}$$

The time the crossing is blocked is now found from $\Delta x = \overline{v}\Delta t$. Note that since the length of the train is given in units of meters, and the units of velocity are in km/h, a conversion of units must be performed in the calculation:

$$\Delta t = \frac{\Delta x}{\overline{v}} = \left(\frac{400 \text{ m}}{49.4 \text{ km/h}}\right)\left(\frac{3600 \text{ s/h}}{1000 \text{ m/km}}\right) = 29.1 \text{ s} \qquad \Diamond$$

47. A small mailbag is released from a helicopter that is descending steadily at 1.50 m/s. After 2.00 s, (a) what is the speed of the mailbag, and (b) how far is it below the helicopter? (c) What are your answers to parts (a) and (b) if the helicopter is rising steadily at 1.50 m/s?

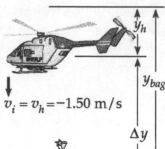

$$v_i = v_h = -1.50 \text{ m/s}$$

Solution

(a) If we choose $+y$ to point upwards, and down as the negative direction, the initial velocity of the mailbag is
$$v_i = v_h = -1.50 \text{ m/s}.$$

As soon as the bag is released, it becomes a freely falling body with acceleration $a = -g = -9.80 \text{ m/s}^2$; the velocity of the bag 2.00 s after it is released is:

$$v_{\text{bag}} = v_i + at = -1.50 \text{ m/s} + \left(-9.80 \text{ m/s}^2\right)(2.00 \text{ s}) = -21.1 \text{ m/s}$$

The **speed** (magnitude of velocity) is: $\text{speed} = |v_{\text{bag}}| = 21.1 \text{ m/s}$ ◊

(b) The displacement of the bag and the helicopter from the release point after 2.00 s are:

$$y_{\text{bag}} = v_i t + \frac{1}{2}at^2 = (-1.50 \text{ m/s})(2.00 \text{ s}) + \frac{1}{2}\left(-9.80 \text{ m/s}^2\right)(2.00 \text{ s})^2 = -22.6 \text{ m}$$

$$y_h = v_h t = (-1.50 \text{ m/s})(2.00 \text{ s}) = -3.00 \text{ m}$$

The difference between them, $\Delta y = y_{\text{bag}} - y_h = \left[-22.6 \text{ m} - (-3.00 \text{ m})\right] = -19.6 \text{ m}$,

and the bag is 19.6 m below the helicopter 2.00 s after it was released.   ◊

(c) If the helicopter (and bag) was moving **upward** at the instant of release, then $v_i = v_h = +1.50 \text{ m/s}$

Using this value for v_i in parts (a) and (b), $v_{\text{bag}} = -18.1 \text{ m/s}$

and $\text{speed} = |v_{\text{bag}}| = 18.1 \text{ m/s}$ ◊

The net displacement of the helicopter and the bag
are $y_{\text{bag}} = -16.6 \text{ m}$ and $y_h = +3.00 \text{ m}$

with a distance between the two of $\Delta y = y_{\text{bag}} - y_h = -19.6 \text{ m}$

Thus, the bag is again 19.6 m below the helicopter 2.00 s after its release.   ◊

51. A student throws a set of keys vertically upward to her sorority sister, who is in a window 4.00 m above. The keys are caught 1.50 s later by the sister's outstretched hand. (a) With what initial velocity were the keys thrown? (b) What was the velocity of the keys just before they were caught?

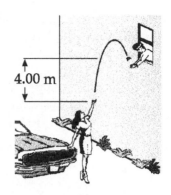

4.00 m

Solution Taking upward as the positive vertical direction, the keys have a constant acceleration of $a = -9.80 \text{ m}/\text{s}^2$ and undergo an upward displacement of $\Delta y = +4.00$ m during the 1.50-s time interval.

(a)   Thus, $\Delta y = v_i t + \frac{1}{2}at^2$ gives the velocity at the beginning of this interval as

$$v_i = \frac{\Delta y - \frac{1}{2}at^2}{t} = \frac{4.00 \text{ m} - \frac{1}{2}\left(-9.80 \text{ m}/\text{s}^2\right)(1.50 \text{ s})^2}{1.50 \text{ s}} = +10.0 \text{ m}/\text{s}$$

$v_i = 10.0 \text{ m}/\text{s}$ **upward**   ◊

(b)   The velocity of the keys 1.50 s later (i.e., just before they are caught) is

$$v_f = v_i + at = +10.0 \text{ m}/\text{s} + \left(-9.80 \text{ m}/\text{s}^2\right)(1.50 \text{ s}) = -4.68 \text{ m}/\text{s}$$

or $v_f = 4.68 \text{ m}/\text{s}$ **downward**   ◊

---

58. A ranger in a national park is driving at 35.0 mi/h when a deer jumps into the road 200 ft ahead of the vehicle. After a reaction time of t, the ranger applies the brakes to produce an acceleration of $a = -9.00 \text{ ft}/\text{s}^2$. What is the maximum reaction time allowed if she is to avoid hitting the deer?

Solution

The equation for uniformly accelerated motion $v_f^2 = v_i^2 + 2a(\Delta x)$ gives the distance required to stop the car once the brakes are applied as

$$(\Delta x)_{\text{stop}} = \frac{v_f^2 - v_i^2}{2a} = \frac{0 - \left[(35 \text{ mi}/\text{h})\left(\dfrac{1.47 \text{ ft}/\text{s}}{1 \text{ mi}/\text{h}}\right)\right]^2}{2\left(-9.00 \text{ ft}/\text{s}^2\right)} = 147 \text{ ft}$$

Thus, if a collision is to be avoided, the maximum distance the car can travel before the brakes are applied is

$$(\Delta x)_{\text{before}} = 200 \text{ ft} - (\Delta x)_{\text{stop}} = 200 \text{ ft} - 147 \text{ ft} = 53.0 \text{ ft}$$

The time required for the car to travel this distance at a constant speed of $v_i = 35.0$ mi / h (and hence the maximum allowed reaction time) is

$$(t_r)_{\text{max}} = \frac{(\Delta x)_{\text{before}}}{v_i} = \frac{53.0 \text{ ft}}{\left[(35.0 \text{ mi / h})\left(\dfrac{1.47 \text{ ft / s}}{1 \text{ mi / h}}\right)\right]} = 1.03 \text{ s} \qquad \Diamond$$

64. In Bosnia, the ultimate test of a young man's courage once was to jump off a 400-year-old bridge (now destroyed) into the River Neretva, 23 m below the bridge. (a) How long did the jump last? (b) How fast was the diver traveling upon impact with the river? (c) If the speed of sound in air is 340 m/s, how long after the diver took off did a spectator on the bridge hear the splash?

Solution

(a) We choose the origins of position and time ($y = 0$ and $t = 0$) to coincide with the diver leaving the bridge. Also, we consider the velocity of the diver as he leaves the bridge to be negligible ($v_i = 0$). The diver is a freely falling body ($a = -g = -9.80 \text{ m / s}^2$) while in the air, and his displacement is related to the elapsed time by $\Delta y = v_i t + \frac{1}{2}at^2$.

The time when he reaches the water, $\qquad\qquad \Delta y = -23.0 \text{ m}$,

is then given by $\qquad\qquad\qquad\qquad\qquad -23 \text{ m} = 0 + \frac{1}{2}\left(-9.80 \text{ m / s}^2\right)t^2$

which reduces to: $\qquad\qquad\qquad\qquad\qquad t = \sqrt{\dfrac{-23 \text{ m}}{-4.90 \text{ m / s}^2}} = 2.2 \text{ s} \qquad \Diamond$

(b) The velocity of the diver as he reaches the water (at $t = 2.2$ s) may be found from

$v_f = v_i + at$ as $\qquad v_f = 0 + \left(-9.80 \text{ m / s}^2\right)(2.2 \text{ s}) = -22 \text{ m / s}$

so the speed (magnitude of velocity) at this time is 22 m/s. $\qquad\qquad\qquad\qquad \Diamond$

(c)   After impact, the time for the sound of the splash to travel (at a constant velocity of 340 m/s) back to a spectator on the bridge

is given by

$$t_{sound} = \frac{\Delta y}{v_{sound}} = \frac{23 \text{ m}}{340 \text{ m / s}} = 0.068 \text{ s}$$

The time between start of the dive and reception of the sound is

$$t_{total} = t + t_{sound} = 2.2 \text{ s} + 0.068 \text{ s} = 2.3 \text{ s} \qquad \Diamond$$

---

67. A stunt woman sitting on a tree limb wishes to drop vertically onto a horse galloping under the tree. The constant speed of the horse is 10.0 m/s, and the woman is initially 3.00 m above the level of the saddle. (a) What must be the horizontal distance between the saddle and limb when the woman makes her move? (b) How long is she in the air?

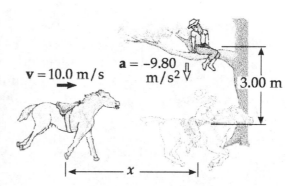

Solution

The most convenient way to solve this problem is to first solve part (b) and use that answer in the solution of part (a) as shown below.

(b)   The stunt woman starts from rest ($v_i = 0$) and is a freely falling body $\left(a = -g = -9.80 \text{ m / s}^2\right)$ until she reaches the saddle. The time to make the 3.00-m vertical drop may be found from $\Delta y = v_i t + \frac{1}{2}at^2$.

This gives

$$-3.00 \text{ m} = 0 + \frac{1}{2}\left(-9.80 \text{ m / s}^2\right)t^2$$

or

$$t = \sqrt{\frac{-3.00 \text{ m}}{-4.90 \text{ m/s}^2}} = 0.782 \text{ s} \qquad \Diamond$$

(a)   Because the horse moves with constant velocity, the horizontal distance it travels during the 0.782 s the stunt woman is falling (and therefore the horizontal distance that should exist between her and the saddle when she makes her move) is given by:

$$\Delta x = v_{horse}t = (10.0 \text{ m / s})(0.782 \text{ s}) = 7.82 \text{ m} \qquad \Diamond$$

Chapter 3

VECTORS AND
TWO-DIMENSIONAL MOTION

NOTES ON SELECTED CHAPTER SECTIONS

3.1 Vectors and Scalars Revisited

A **scalar** has only magnitude and no direction. On the other hand, a **vector** is a physical quantity that requires the specification of both direction and magnitude.

3.2 Some Properties of Vectors

Two vectors are **equal** if they have both the same magnitude and direction. When two or more vectors are **added**, they must have the same units. Two vectors which are the **negative** of each other have the same magnitude but opposite directions. When a vector is **multiplied** by a positive (negative) scalar, the result is a vector in the same (opposite) direction. The magnitude of the resulting vector is equal to the product of the scalar and the magnitude of the original vector.

3.4 Displacement, Velocity, and Acceleration in Two Dimensions

Consider an object moving in the plane containing the x and y axes. The **displacement** of the object is defined as the change in the position vector, $\Delta\mathbf{r}$. The **average velocity** of a particle during the time interval Δt is the ratio of the displacement to the time interval for this displacement. The average velocity is a **vector** quantity directed along $\Delta\mathbf{r}$. The **instantaneous velocity v** is defined as the limit of the average velocity, $\Delta\mathbf{r}/\Delta t$, as Δt goes to zero. The direction of the instantaneous velocity vector is along a line that is tangent to the path of the particle and in the direction of motion. The **average acceleration** of an object whose velocity changes is the ratio of the net change in velocity to the time interval during which the change occurs, $\Delta\mathbf{v}/\Delta t$.

A particle can accelerate in several ways: the magnitude of the velocity vector (the speed) may change; the direction of the velocity vector may change, making a curved path, even though the speed is constant; or both the magnitude and direction of the velocity vector may change.

3.5 Projectile Motion

In the case of projectile motion, if it is assumed that air resistance is negligible and that the rotation of the Earth does not affect the motion, then:

1. the horizontal component of velocity, v_x, remains constant because there is no horizontal component of acceleration;

2. the vertical component of acceleration is equal to the acceleration due to gravity, **g**;

3. the vertical component of velocity, v_y, and the displacement in the y direction are identical to those of a freely falling body;

4. projectile motion can be described as a superposition of the two motions in the x and y directions.

Review the procedure recommended in the Suggestions, Skills, and Strategies section for solving projectile motion problems.

3.6 Relative Velocity

Observations made by observers in different frames of reference can be related to one another through the techniques of the transformation of relative velocities.

When two objects are each moving with respect to a stationary reference frame (e.g. the Earth), each moving object has a **relative velocity with respect to the other one**. You must use **vector addition** in order to determine the relative velocity of one object with respect to the other one. Start by drawing a **vector diagram** showing the way in which the magnitude and direction of the velocities (object "one", object "two", and the fixed reference) are related to each other. There is no general equation which can be memorized and used to make relative velocity calculations. Refer to your vector diagram and identify each velocity vector with subscripts as illustrated in Example 3.7 and Example 3.8 in your textbook.

EQUATIONS AND CONCEPTS

Vector quantities obey the commutative law of addition. In order to add vector **A** to vector **B** using the graphical method, first construct **A**, and then draw **B** such that the tail of **B** starts at the head of **A**. The sum of **A** + **B** is the vector that completes the triangle by connecting the tail of **A** to the head of **B**.

$$A + B = B + A$$

When more than two vectors are to be added, they are all connected head-to-tail in any order and the resultant or sum is the vector which joins the tail of the first vector to the head of the last vector.

$$R = A + B + C + D$$

When two or more vectors are to be added, all of them must represent the same physical quantity—that is, have the same units. In the graphical or geometric method of vector addition, **the length of each vector corresponds to the magnitude of the vector according to a chosen scale.** Also, the direction of each vector must be along a direction which makes the proper angle relative to the others.

Comment on
vector addition

The operation of vector subtraction utilizes the definition of the negative of a vector. The negative of vector **A** is the vector which has a magnitude equal to the magnitude of **A**, but acts or points along a direction opposite the direction of **A**.

$$A - B = A + (-B) \tag{3.1}$$

A vector **A** in a two-dimensional coordinate system can be resolved into its components along the x and y directions. The projection of **A** onto the x axis, A_x, is the x component of **A**; and the projection of **A** onto the y axis, A_y, is the y component of **A**.

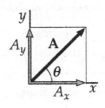

The magnitude of **A** and the angle θ, which the vector makes with the positive x-axis, can be determined from the values of the x and y components of **A**.

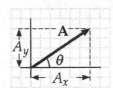

$$A_x = A\cos\theta \tag{3.2}$$

$$A_y = A\sin\theta$$

$$A = \sqrt{A_x{}^2 + A_y{}^2} \tag{3.3}$$

$$\tan\theta = \frac{A_y}{A_x} \tag{3.4}$$

A projectile moves in two directions (e.g. the x and y directions) simultaneously. The acceleration along the vertical direction is $-g$ (as in free fall) and the acceleration along the horizontal direction is zero (when air resistance is neglected).

The path of a projectile is curved as shown in the figure to the right. Such a curve is called a parabola. The vector which represents the initial velocity, v_0, makes an angle of θ_0 with the horizontal where θ_0 is called the projection angle or angle of launch. In order to analyze projectile motion, you should separate the motion into two parts, the x (horizontal) motion and the y (vertical) motion, and apply the equations of constant acceleration to each part separately.

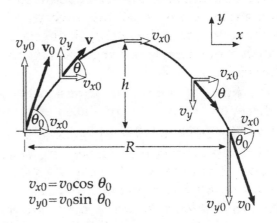

$$v_{x0} = v_0\cos\theta_0$$
$$v_{y0} = v_0\sin\theta_0$$

Velocity components
in projectile motion

The initial horizontal and vertical components of velocity of a projectile depend on the magnitude of the initial velocity vector and the initial angle of launch.

$$v_{x0} = v_0 \cos \theta_0$$

$$v_{y0} = v_0 \sin \theta_0$$

The horizontal component of velocity for a projectile remains constant ($a_x = 0$); while the vertical component decreases uniformly with time ($a_y = -g$). Note the changing value of v_y and constant value of v_x as shown in the figure at the bottom of the previous page.

$$v_x = v_{x0} = v_0 \cos \theta_0 = \text{constant} \qquad (3.9)$$

$$v_y = v_{y0} - gt \qquad (3.11)$$

or

$$v_y = v_0 \sin \theta_0 - gt$$

The x and y coordinates of the position of a projectile are functions of the elapsed time.

$$\Delta x = v_{x0} t = (v_0 \cos \theta_0)t \qquad (3.10)$$

The positive direction for the vertical motion is assumed to be upward.

$$\Delta y = v_{y0} t - \frac{1}{2} gt^2 \qquad (3.12)$$

or

$$\Delta y = (v_0 \sin \theta_0)t - \frac{1}{2} gt^2$$

The initial and final velocities in the y direction and the displacement are related as stated in Eq. 3.13.

$$v_y{}^2 = v_{y0}{}^2 - 2g\Delta y \qquad (3.13)$$

If v_x and v_y are known at any point, the angle that the velocity vector makes with the x-axis can be determined.

$$\theta = \tan^{-1}\left(\frac{v_y}{v_x}\right)$$

SUGGESTIONS, SKILLS, AND STRATEGIES

ADDITION AND SUBTRACTION OF VECTORS

When two or more vectors are to be added, the following step-by-step procedure is recommended:

1. Select a coordinate system.

2. Draw a sketch of the vectors to be added (or subtracted), with a label on each vector.

3. Find the x and y components of all vectors.

4. Find the algebraic sum of the x-components of the vectors, and the algebraic sum of the y-components of the vectors. These two sums are the **x and y components of the resultant vector**.

5. Use the Pythagorean theorem to find the magnitude of the resultant vector.

6. Use a suitable trigonometric function (e.g. Equation 3.4) to find the angle the resultant vector makes with the x axis.

PROJECTILE MOTION PROBLEMS

The following procedure is recommended for solving projectile motion problems:

1. Sketch the path of the projectile on a set of coordinate axes. Include the initial velocity vector and the projectile angle.

2. Resolve the initial velocity vector into x and y components.

3. Treat the horizontal motion (zero acceleration) and the vertical motion (acceleration = $-g$) independently.

4. Follow the techniques for solving problems with constant velocity (zero acceleration) to analyze the horizontal motion of the projectile.

5. Follow the techniques for solving problems with constant acceleration to analyze the vertical motion of the projectile.

THE PATH, MAXIMUM HEIGHT, AND RANGE OF A PROJECTILE

The equation for the path (trajectory) of a projectile can be found by combining Equation 3.10, $x = (v_0 \cos\theta_0)t$, and the equation following Equation 3.12, $y = (v_0 \sin\theta_0)t - \frac{1}{2}gt^2$. Solve Equation 3.10 for t to get $t = x/(v_0 \cos\theta_0)$, and substitute this expression for t into the equation for y to find

$$y = (v_0 \sin\theta_0)\frac{x}{v_0 \cos\theta_0} - \frac{1}{2}g\left(\frac{x}{v_0 \cos\theta_0}\right)^2$$

or

$$y = (\tan\theta_0)x - \left(\frac{g}{2v_0{}^2 \cos^2\theta_0}\right)x^2$$

Note that this is in the form of the **equation of a parabola**.

When analyzing projectile motion, there are two quantities of particular interest. These are the horizontal range, R (the maximum value of x), and the maximum height, H (the maximum value of y). These quantities can be easily determined for a **projectile that impacts at the same level from which it was launched**.

FINDING THE MAXIMUM HEIGHT, H

When the projectile reaches maximum height the y-component of the velocity will be zero; otherwise, the projectile would be continuing to rise or would be falling. Therefore, use the equation $v_y = v_0 \sin\theta_0 - gt$ with $v_y = 0$ and $t = t_1$, the time it takes the projectile to reach maximum height. Solve $0 = v_0 \sin\theta_0 - gt_1$ for the time it takes the projectile to reach the peak of its curve:

$$t_1 = \frac{v_0 \sin\theta_0}{g}$$

Then, noting that when $t = t_1$, $\qquad\qquad y = H$

substitute both these values into $\qquad y = (v_0 \sin\theta_0)t - \frac{1}{2}gt^2$

to get $\qquad\qquad\qquad\qquad H = (v_0 \sin\theta_0)\left(\frac{v_0 \sin\theta_0}{g}\right) - \frac{1}{2}g\left(\frac{v_0 \sin\theta_0}{g}\right)^2$

Simplifying to find the maximum height, $\quad H = \dfrac{v_0{}^2 \sin^2\theta_0}{2g}$

FINDING THE RANGE, R

When the projectile **lands on the same level from which it was launched,** $R = x$ when $y = 0$. From the equation of the path (trajectory equation),

$$y = (\tan\theta_0)x - \left(\frac{g}{2v_0^2 \cos^2\theta_0}\right)x^2 \quad \text{when} \quad y = 0$$

Then

$$x = R = \frac{2v_0^2 \cos^2\theta_0 \tan\theta_0}{g} = \frac{2v_0^2 \sin\theta_0 \cos\theta_0}{g}$$

Finally, use the trigonometric identity $\quad \sin 2\theta = 2\sin\theta\cos\theta$

to simplify the expression for **range** $\quad R = \frac{v_0^2}{g}\sin 2\theta_0$

REVIEW CHECKLIST

▷ Recognize that two-dimensional motion in the xy plane with constant acceleration is equivalent to two independent motions: constant velocity along the x-direction and constant acceleration along the y-direction. (Section 3.5)

▷ Sketch a typical trajectory of a particle moving in the xy plane and draw vectors to illustrate the manner in which the displacement, velocity, and acceleration of the particle changes with time. (Section 3.4)

▷ Recognize the fact that if the initial speed v_0 and initial angle θ_0 of a projectile are known at a given point at $t = 0$, the velocity components and coordinates can be found at any later time t. Furthermore, one can also calculate the horizontal range R and maximum height H if v_0 and θ_0 are known. (Section 3.5)

▷ Understand and describe the basic properties of vectors, resolve a vector into its rectangular components, use the rules of vector addition (including graphical solutions for addition of two or more vectors), and determine the magnitude and direction of a vector from its rectangular components. (Sections 3.2 and 3.3)

▷ Practice the technique demonstrated in the text to solve relative velocity problems. (Section 3.6)

SOLUTIONS TO SELECTED END-OF-CHAPTER PROBLEMS

4. A jogger runs 100 m due west, then changes direction for the second leg of the run. At the end of the run, she is 175 m away from the starting point at an angle of 15.0° north of west. What were the direction and length of her second displacement? Use graphical techniques.

Solution The vector **R** representing the jogger's net displacement from the starting point to the end of the run is

$$\mathbf{R} = \mathbf{A} + \mathbf{B}$$

where **A** and **B** are the displacements that occur during each of the two legs of the run.

This equation may also be written as

$$\mathbf{B} = \mathbf{R} - \mathbf{A} = \mathbf{R} + (-\mathbf{A})$$

Thus, we can solve for the second displacement **B** by using graphical techniques to add the known resultant, **R** = 175 m at 15.0° north of west, to the vector −**A**.

The negative of **A** is a vector having the same magnitude as **A** but whose direction is opposite that of **A**.

Therefore,

$$-\mathbf{A} = 100 \text{ m due east}$$

The sketch (above) shows a scale drawing you can make to solve for the second displacement. After the known vectors **R** and −**A** are drawn to the proper length (according to your chosen scale) and in the specified directions, the vector representing their sum is that drawn from the start of **R** to the end of −**A** as shown. The magnitude of **B** is found by measuring the length of this vector and multiplying by your scale factor. The angle θ gives the direction of **B** and may be measured on your scale drawing using a protractor.

You should find that **B** has a magnitude of   approximately 83 m

and is oriented at   about 33° north of west.   ◊

11. A girl delivering newspapers covers her route by traveling 3.00 blocks west, 4.00 blocks north, then 6.00 blocks east. (a) What is her resultant displacement? (b) What is the total distance she travels?

Solution

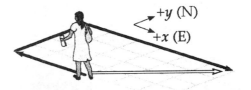

(a) If eastward and northward are chosen as $+x$ and the $+y$ directions respectively, the components of each displacement are:

Displacement	x-component	y-component
A	−3.00 blocks	0.00
B	0.00	+4.00 blocks
C	+6.00 blocks	0.00

The components of the resultant are then:

$$R_x = A_x + B_x + C_x = +3.00 \text{ blocks}$$

$$R_y = A_y + B_y + C_y = +4.00 \text{ blocks}$$

The magnitude and direction of the resultant are then given by:

$$R = \sqrt{R_x^{\,2} + R_y^{\,2}} = 5.00 \text{ blocks} \qquad \Diamond$$

$$\theta = \tan^{-1}\left(\frac{R_y}{R_x}\right) = 53.1° \text{ north of east} \qquad \Diamond$$

(b) The distance traveled is $\quad d = (3.00 + 4.00 + 6.00) \text{ blocks} = 13.0 \text{ blocks} \qquad \Diamond$

19. A man pushing a mop across a floor causes the mop to undergo two displacements. The first has a magnitude of 150 cm and makes an angle of 120° with the positive x axis. The resultant displacement has a magnitude of 140 cm and is directed at an angle of 35.0° to the positive x axis. Find the magnitude and direction of the second displacement.

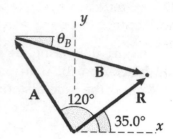

Solution If **A** and **B** are the first and second displacements, the resultant displacement is

$$\mathbf{R} = \mathbf{A} + \mathbf{B}$$

If we take A_x, A_y, B_x, and B_y to be the components of the first and second displacements, the components of the resultant displacement are given by

$$R_x = A_x + B_x \quad \text{and} \quad R_y = A_y + B_y$$

The vectors **R** and **A** are known. Their components are as follows:

x-components:
$$R_x = R\cos\theta_R = (140 \text{ cm})\cos 35.0° = +115 \text{ cm}$$

$$A_x = A\cos\theta_A = (150 \text{ cm})\cos 120° = -75.0 \text{ cm}$$

y-components:
$$R_y = R\sin\theta_R = (140 \text{ cm})\sin 35.0° = +80.3 \text{ cm}$$

$$A_y = A\sin\theta_A = (150 \text{ cm})\sin 120° = +130 \text{ cm}$$

The components of the second displacement **B** may then be found to be:

$$B_x = R_x - A_x = 190 \text{ cm}$$

$$B_y = R_y - A_y = -49.7 \text{ cm}$$

The magnitude of **B** is $B = \sqrt{B_x{}^2 + B_y{}^2} = \sqrt{(190 \text{ cm})^2 + (-49.7 \text{ cm})^2} = 196 \text{ cm}$ ◊

The direction of **B** is $\theta_B = \tan^{-1}\left(\dfrac{B_y}{B_x}\right) = \tan^{-1}(-0.261) = -14.7°$ ◊

Thus, $\mathbf{B} = 196 \text{ cm}$ at 14.7° below the positive x direction.   ◊

27. A tennis player standing 12.6 m from the net hits the ball at 3.00° above the horizontal. To clear the net, the ball must rise at least 0.330 m. If the ball just clears the net at the apex of its trajectory, how fast was the ball moving when it left the racquet?

Solution

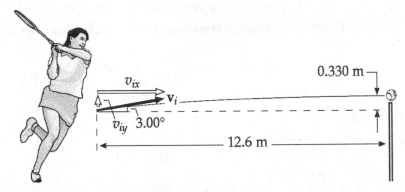

At the apex of the trajectory, $v_y = 0$. Therefore, $v_y = v_{iy} + a_y t$ gives the time to reach the net as

$$t = \frac{v_y - v_{iy}}{a_y} = \frac{0 - v_i \sin 3.00°}{-g} = \frac{v_i \sin 3.00°}{g}$$

Since the vertical acceleration is constant, the average velocity in the vertical direction is

$$\bar{v}_y = \frac{v_y + v_{iy}}{2}$$

For the time interval from when the ball leaves the racquet until when it reaches the net, this becomes

$$\bar{v}_y = \frac{0 + v_i \sin 3.00°}{2} = \frac{v_i \sin 3.00°}{2}$$

The vertical displacement during this time interval is

$$\Delta y = \bar{v}_y t = \left(\frac{v_i \sin 3.00°}{2} \right)\left(\frac{v_i \sin 3.00°}{g} \right) = \frac{v_i^2 \sin^2 3.00°}{2g}$$

If the ball just clears the net, then $\Delta y = 0.330$ m, giving

$$v_i = \frac{\sqrt{2g(\Delta y)}}{\sin 3.00°} = \frac{\sqrt{2(9.80 \text{ m} / \text{s}^2)(0.330 \text{ m})}}{\sin 3.00°} = 48.6 \text{ m} / \text{s} \qquad \Diamond$$

Note that it was unnecessary to use the horizontal distance of 12.6 m in this solution.

31. A car is parked on a cliff overlooking the ocean on an incline that makes an angle of 24.0° below the horizontal. The negligent driver leaves the car in neutral, and the emergency brakes are defective. The car rolls from rest down the incline with a constant acceleration of 4.00 m / s² for a distance of 50.0 m to the edge of the cliff. The cliff is 30.0 m above the ocean. Find (a) the car's position relative to the base of the cliff when the car lands in the ocean, and (b) the length of time the car is in the air.

Solution   First, consider the motion of the car down the incline (a uniformly accelerated motion in one dimension). Using $v^2 = v_i^2 + 2a(\Delta s)$, the speed of the car when it reaches the edge of the cliff is found to be

$$v = \sqrt{v_i^2 + 2a(\Delta s)} = \sqrt{0 + 2(4.00 \text{ m / s}^2)(50.0 \text{ m})} = 20.0 \text{ m / s}$$

Therefore, at the instant the car becomes a projectile, its velocity components are

$$v_{ix} = v(\cos 24.0°) = 18.3 \text{ m / s} \qquad \text{and} \qquad v_{iy} = -v(\sin 24.0°) = -8.13 \text{ m / s}$$

During the projectile phase of the car's motion, its acceleration components are

$$a_x = 0 \qquad \text{and} \qquad a_y = -g = -9.80 \text{ m / s}^2$$

(b)   As the car drops from the edge of the cliff to the ocean, the vertical displacement is $\Delta y = -30.0 \text{ m}$. Thus, the vertical velocity when it reaches ocean level is

$$v_y = -\sqrt{v_{iy}^2 + 2a_y(\Delta y)} = \sqrt{(-8.13 \text{ m / s})^2 + 2(-9.80 \text{ m / s}^2)(-30.0 \text{ m})} = -25.6 \text{ m / s}$$

The time the car takes to fall from the cliff top to the ocean is

$$t = \frac{v_y - v_{iy}}{a_y} = \frac{-(25.6 \text{ m / s}) - (-8.13 \text{ m / s})}{-9.80 \text{ m / s}^2} = 1.78 \text{ s}$$

(a)   When the car reaches ocean level, the horizontal distance between it and the cliff is the horizontal displacement occurring during the time found above. This distance is

$$\Delta x = v_{ix}t + \frac{1}{2}a_x t^2 = (18.3 \text{ m / s})(1.78 \text{ s}) + 0 = 32.5 \text{ m}$$

◊

39. A rowboat crosses a river with a velocity of 3.30 mi/h at an angle 62.5° north of west relative to the water. The river is 0.505 mi wide and carries an eastward current of 1.25 mi/h. How far upstream is the boat when it reaches the opposite shore?

Solution

The velocity of the boat relative to the shore, v_{bs}, may be expressed as the sum $v_{bs} = v_{bw} + v_{ws}$, where v_{bw} is the velocity of the boat relative to the water and v_{ws} is the velocity of the water relative to the shore.

Graphical addition of these vectors is illustrated in the sketch to the right.

Each component of the boat's velocity is as follows:

Directed across the stream (northward),

$$(v_{bs})_{north} = (v_{bw})_{north} + (v_{ws})_{north} = (3.30 \text{ mi / h})\sin 62.5° + 0 = 2.93 \text{ mi / h}$$

Directed parallel to the stream (eastward),

$$(v_{bs})_{east} = (v_{bw})_{east} + (v_{ws})_{east} = -(3.30 \text{ mi / h})\cos 62.5° + 1.25 \text{ mi / h} = -0.274 \text{ mi / h}$$

The time required for the boat to cross the stream (i.e., move 0.505 mi north) is therefore

$$t = \frac{0.505 \text{ mi}}{(v_{bs})_{north}} = \frac{0.505 \text{ mi}}{2.93 \text{ mi / h}} = 0.172 \text{ h}$$

The displacement of the boat parallel to the stream during this time is given by:

$$x = (v_{bs})_{east} t = (-0.274 \text{ mi / h})(0.172 \text{ h}) = -4.71 \times 10^{-2} \text{ mi}$$

Thus, as the boat crosses the river, it moves in the negative eastward (i.e., westward or upstream) direction a distance of

$$|x| = (4.71 \times 10^{-2} \text{ mi})\left(\frac{5280 \text{ ft}}{1.00 \text{ mi}}\right) = 249 \text{ ft}$$ ◊

47. Towns A and B in Figure P3.47 are 80.0 km apart. A couple arranges to drive from town A and meet a couple driving from town B at the lake, L. The two couples leave simultaneously and drive for 2.50 h in the directions shown. Car 1 has a speed of 90.0 km/h. If the cars arrive simultaneously at the lake, what is the speed of car 2?

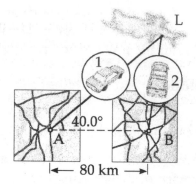
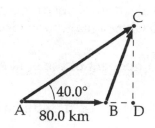

Figure P3.47

Solution

The vector diagram at the right shows the displacement between the cities as **AB** (magnitude $\overline{AB} = 80.0$ km), the displacement undergone by Car 1 as **AC** (magnitude $\overline{AC}$), and the displacement of Car 2 as **BC** (magnitude $\overline{BC}$).

Observe that

$$\overline{AC} = v_1 t = (90 \text{ km / h})(2.50 \text{ h}) = 225 \text{ km}$$

From the right triangle ADC,

$$\overline{BD} = \overline{AD} - \overline{AB} = \overline{AC}(\cos 40.0°) - 80.0 \text{ km}$$
$$= (225 \text{ km})\cos 40.0° - 80.0 \text{ km} = 92.4 \text{ km}$$

Finally, using triangle BDC, the Pythagorean theorem gives

$$\overline{BC} = \sqrt{\left(\overline{BD}\right)^2 + \left(\overline{DC}\right)^2} = \sqrt{\left(\overline{BD}\right)^2 + \left(\overline{AC}\sin 40.0°\right)^2}$$

or

$$\overline{BC} = \sqrt{(92.4 \text{ km})^2 + \left[(225 \text{ km})\sin 40.0°\right]^2} = 172 \text{ km}$$

Hence, the speed of Car 2 is

$$v_2 = \frac{\overline{BC}}{t} = \frac{172 \text{ km}}{2.50 \text{ h}} = 68.6 \text{ km / h} \qquad \Diamond$$

52. A daredevil decides to jump a canyon. Its walls are equally high and 10 m apart. He takes off by driving a motorcycle up a short ramp sloped at an angle of 15°. What minimum speed must he have in order to clear the canyon?

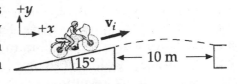

Solution

Choosing the reference axes as shown in the sketch, the components of the initial velocity of the daredevil are

$$v_{iy} = v_i \sin 15° \qquad \text{and} \qquad v_{ix} = v_i \cos 15°$$

His acceleration components are $\quad a_x = 0 \qquad$ and $\qquad a_y = -g = -9.80 \text{ m / s}^2$

The y-coordinate of the daredevil at any time is given by $\qquad y = v_{iy}t + \frac{1}{2}a_y t^2$

When the jumper is at the level of the canyon rim ($y = 0$), this equation reduces to

$$0 = \left(v_i \sin 15°\right)t - \left(4.90 \text{ m / s}^2\right)t^2$$

This has one solution, $t = 0$, which coincides with the start of the jump, and a second solution, $t = \left(v_i \sin 15° / 4.90 \text{ m / s}^2\right)$, which gives the time when the jumper returns to the level of the canyon rim.

Since $a_x = 0$, the jumper maintains a constant horizontal velocity $v_x = v_{ix}$, and his horizontal displacement at any time is given by $x = v_{ix}t$. If the jumper is to successfully cross the canyon, it is necessary to have $x \geq 10$ m at the time he returns to the level of the rim. Thus, the requirement for a successful jump becomes

$$x = v_{ix}t = \left(v_i \cos 15°\right)\left(\frac{v_i \sin 15°}{4.90 \text{ m / s}^2}\right) \geq 10 \text{ m}$$

with an initial speed $\qquad v_i \geq \sqrt{\dfrac{(10 \text{ m})\left(4.90 \text{ m / s}^2\right)}{(\sin 15°)(\cos 15°)}} = 14 \text{ m / s} \qquad \diamond$

55. A home run is hit in such a way that the baseball just clears a wall 21m high, located 130 m from home plate. The ball is hit at an angle of 35° to the horizontal, and air resistance is negligible. Find (a) the initial speed of the ball, (b) the time it takes the ball to reach the wall, and (c) the velocity components and the speed of the ball when it reaches the wall. (Assume the ball is hit at a height of 1.0 m above the ground.)

Solution

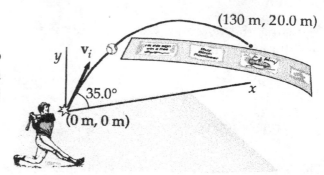

(a) The time required for the ball to reach the wall (i.e., achieve a horizontal displacement of 130 m) is

$$t = \frac{\Delta x}{v_{ix}} = \frac{130 \text{ m}}{v_i \cos 35°} = \frac{159 \text{ m}}{v_i}$$

At this time, the ball must be 21 m above the ground, or 20 m above its launch point ($\Delta y = +20$ m).

Therefore, $\quad \Delta y = v_{iy} t + \frac{1}{2} a_y t^2$

becomes $\quad 20 \text{ m} = (v_i \sin 35°)\left(\frac{159 \text{ m}}{v_i}\right) + \frac{1}{2}(-9.80 \text{ m} / \text{s}^2)\left(\frac{159 \text{ m}}{v_i}\right)^2$

Simplifying and solving for the initial velocity gives $\quad v_i = 42 \text{ m} / \text{s}$ ◊

(b) From above, the elapsed time when the ball reaches the wall is

$$t = \frac{159 \text{ m}}{v_i} = \frac{159 \text{ m}}{42 \text{ m} / \text{s}} = 3.8 \text{ s}$$ ◊

(c) At this time, the velocity components of the ball are

$$v_x = v_{ix} = v_i \cos 35° = (42 \text{ m} / \text{s}) \cos 35° = 34 \text{ m} / \text{s}$$ ◊

and $\quad v_y = v_{iy} + a_y t = (42 \text{ m} / \text{s}) \sin 35° + (-9.80 \text{ m} / \text{s}^2)(3.8 \text{ s}) = -13 \text{ m} / \text{s}$ ◊

The speed of the ball as it crosses the wall is $\quad v = \sqrt{v_x^2 + v_y^2} = 37 \text{ m} / \text{s}$ ◊

63. A hunter wishes to cross a river that is 1.5 km wide and flows with a speed of 5.0 km/h parallel to its banks. The hunter uses a small powerboat that moves at a maximum speed of 12 km/h with respect to the water. What is the minimum time necessary for crossing?

Solution

The hunter will cross the river in minimum time if his velocity relative to the water carries him perpendicularly to the flow of the stream (i.e., straight across the current). The velocity of the hunter relative to the earth equals the vector sum of his velocity relative to the water and the velocity of the water relative to the earth.

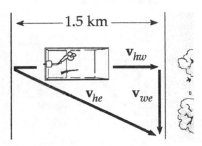

This is summarized in the equation $\mathbf{v}_{he} = \mathbf{v}_{hw} + \mathbf{v}_{we}$

and the vector triangle shown in the sketch illustrates the conditions for minimum crossing time. Under these conditions,

the time to cross the river is given by $t = \dfrac{\text{width of stream}}{v_{hw}}$

Substituting known values, $t = \dfrac{(1.5\ \text{km})(60\ \text{min}/\text{h})}{12\ \text{km}/\text{h}} = 7.5\ \text{min}$ ◊

Also, during this time the hunter will be carried downstream a distance of

$$d = v_{we}t = \dfrac{(5.0\ \text{km}/\text{h})(7.5\ \text{min})}{60\ \text{min}/\text{h}} = 0.63\ \text{km}$$

41

65. A daredevil is shot out of a cannon at 45.0° to the horizontal with an initial speed of 25.0 m/s. A net is positioned a horizontal distance of 50.0 m from the cannon. At what height above the cannon should the net be placed in order to catch the daredevil?

Solution

Choose a reference frame with its origin at the point where the daredevil leaves the cannon, with the x-axis horizontal and the y-axis vertical.

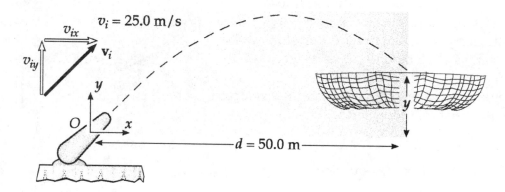

Then, the components of the daredevil's initial velocity and acceleration are:

$$v_{ix} = (25.0 \text{ m / s}) \cos 45° = 17.7 \text{ m / s} \qquad a_x = 0$$

$$v_{iy} = (25.0 \text{ m / s}) \sin 45° = 17.7 \text{ m / s} \qquad \text{and} \qquad a_y = -g = -9.80 \text{ m / s}^2$$

Thus, the horizontal velocity $v_x = v_{ix} + a_x t$ is constant and the time to travel the horizontal distance of 50.0 m to the net is

$$t = \frac{d}{v_x} = \frac{50.0 \text{ m}}{17.7 \text{ m / s}} = 2.83 \text{ s}$$

The daredevil's y-coordinate at this time is given by $y = v_{iy} t + \frac{1}{2} a_y t^2$

$$y = (17.7 \text{ m / s})(2.83 \text{ s}) + \frac{1}{2}\left(-9.80 \text{ m / s}^2\right)(2.83 \text{ s})^2 = +10.8 \text{ m}$$ ◊

69. Instructions for finding a buried treasure include the following: Go 75.0 paces at 240°, turn to 135° and walk 125 paces, then travel 100 paces at 160°. Determine the resultant displacement from the starting point.

Solution

Choosing the origin of the reference frame at the starting point, the components of the individual displacements are:

$$d_{1x} = (75.0)\cos 240° = -37.5 \text{ paces} \qquad d_{1y} = (75.0)\sin 240° = -65.0 \text{ paces}$$

$$d_{2x} = (125)\cos 135° = -88.4 \text{ paces} \qquad d_{2y} = (125)\sin 135° = +88.4 \text{ paces}$$

$$d_{3x} = (100)\cos 160° = -94.0 \text{ paces} \qquad d_{3y} = (100)\sin 160° = +34.2 \text{ paces}$$

The x- and y-components of the resultant displacement are then given by:

$$R_x = \sum_{i=1}^{3} d_{ix} = (-37.5 - 88.4 - 94.0) \text{ paces} = -220 \text{ paces}$$

and

$$R_y = \sum_{i=1}^{3} d_{iy} = (-65.0 + 88.4 + 34.2) \text{ paces} = +57.6 \text{ paces}$$

The magnitude of the resultant is

$$R = \sqrt{R_x^2 + R_y^2}$$

or

$$R = \sqrt{(-220 \text{ paces})^2 + (+57.6 \text{ paces})^2} = 227 \text{ paces}$$

The direction of the resultant is found from

$$\theta = \tan^{-1}\left(\frac{R_y}{R_x}\right) = \tan^{-1}\left(\frac{+57.6 \text{ paces}}{-220 \text{ paces}}\right) = \tan^{-1}(-0.262)$$

This yields two possible answers: $\theta = -14.7°$, or $\theta = 165°$. When it is observed that $R_x < 0$, while $R_y > 0$, it becomes clear that the resultant vector $\mathbf{R}$ must lie in the second quadrant, not the fourth. The desired angle is then $\theta = 165°$, and the resultant is:

$$\mathbf{R} = 227 \text{ paces at } 165°$$

Chapter 4
THE LAWS OF MOTION

NOTES ON SELECTED CHAPTER SECTIONS

4.1   The Concept of Force

Equilibrium is the condition under which the **net force** (vector sum of all forces) acting on an object is zero. An object in equilibrium has a zero acceleration (velocity is constant or equals zero).

Fundamental forces in nature are:

1. gravitational (attractive forces between objects due to their masses)
2. electromagnetic forces (between electric charges at rest or in motion)
3. strong nuclear forces (between subatomic particles)
4. weak nuclear forces (accompanying the process of radioactive decay).

Classical physics is concerned with contact forces (which are the result of physical contact between two or more objects) and action-at-a-distance forces (which act through empty space and do not involve physical contact).

4.2   Newton's First Law

Newton's first law is called the **law of inertia** and states that an object at rest will remain at rest and an object in motion will remain in motion with a constant velocity unless acted on by a **net external force**.

Mass and **weight** are two different physical quantities. The weight of a body is equal to the **force of gravity** acting on the body and varies with location in the Earth's gravitational field. Mass is an inherent property of a body and is a measure of the body's inertia (resistance to change in its state of motion). The SI unit of mass is the **kilogram** and the unit of weight is the **newton**.

4.3   Newton's Second Law

Newton's second law, the **law of acceleration,** states that the acceleration of an object is directly proportional to the resultant force acting on it and inversely proportional to its mass. The direction of the acceleration is in the direction of the net force.

4.4 Newton's Third Law

Newton's third law, the law of action-reaction, states that when two bodies interact, the force which body "A" exerts on body "B" (the action force) is equal in magnitude and opposite in direction to the force which body "B" exerts on body "A" (the reaction force). A consequence of the third law is that forces occur in pairs. Remember that the action force and the reaction force never cancel because they act on different objects.

4.5 Some Applications of Newton's Laws

Construction of a free-body diagram is an important step in the application of Newton's laws of motion to solve problems involving bodies in equilibrium or accelerating under the action of external forces. The diagram should include a labeled arrow to identify each of the external forces acting on the body whose motion (or condition of equilibrium) is to be studied. Forces which are the reactions to these external forces must not be included. When a system consists of more than one body or mass, you must construct a free-body diagram for each mass.

4.6 Forces of friction

When a body is in motion either on a surface or through a viscous medium such as air or water, there is resistance to the motion because the body interacts with its surroundings. We call such resistance a force of friction. Experiments show that the frictional force arises from the nature of the two surfaces. To a good approximation, both $f_{s,max}$ (maximum force of static friction) and f_k (force of kinetic friction) are proportional to the normal force at the interface between the two surfaces.

EQUATIONS AND CONCEPTS

A quantitative measurement of mass (the term used to measure inertia) can be made by comparing the accelerations that a given force will produce on different bodies. If a given force acting on a body of mass m_1 produces an acceleration a_1 and the same force acting on a body of mass m_2 produces an acceleration a_2, the ratio of the two masses equals the inverse of the ratio of the two accelerations.

$$\frac{m_1}{m_2} = \frac{a_2}{a_1}$$

The acceleration of an object is proportional to the resultant force acting on it and inversely proportional to its mass. This is a statement of Newton's second law.

$$\Sigma \mathbf{F} = m\mathbf{a} \qquad (4.1)$$

When several forces act on an object, it is often convenient to write the vector form of the equation expressing Newton's second law as component scalar equations. The orientation of the coordinate system can often be chosen so that the object has a nonzero acceleration along only one direction.

$$\Sigma F_x = ma_x \qquad (4.2)$$
$$\Sigma F_y = ma_y$$
$$\Sigma F_z = ma_z$$

Calculations with Equations 4.1 and 4.2 must be made using a consistent set of units for the quantities force, mass, and acceleration. The SI unit of force is the newton (N), defined as the force that, when acting on a 1-kg mass, produces an acceleration of 1 m/s².

$$1\,\mathrm{N} \equiv 1\,\mathrm{kg \cdot m / s^2} \qquad (4.3)$$

$$1\,\mathrm{dyne} \equiv 1\,\mathrm{g \cdot cm / s^2}$$

$$1\,\mathrm{N} \equiv 0.225\,\mathrm{lb} \qquad (4.4)$$

Newton's **law of universal gravitation** states that every particle in the universe attracts every other particle with a force that is directly proportional to the product of the masses of the particles and inversely proportional to the square of the distance between them.

$$F_g = G\frac{m_1 m_2}{r^2} \qquad (4.5)$$

The **universal gravitational constant**.

$$G = 6.67 \times 10^{-11}\,\mathrm{N \cdot m^2 / kg^2}$$

Weight is not an inherent property of a body, but depends on the local value of g and varies with location.

$$w = mg \qquad (4.6)$$

The value of the acceleration due to gravity, g, decreases with increasing distance from the center of the Earth.

$$g = G\frac{M_E}{r^2} \qquad (4.8)$$

This is a statement of **Newton's third law**, which states that forces always occur in pairs; and the force exerted by body 1 on body 2 (the action force) is equal in magnitude and opposite in direction to the force exerted by body 2 on body 1 (the reaction force). **The action and reaction forces always act on different objects.**

$$\mathbf{F}_{12} = -\mathbf{F}_{21}$$

Equilibrium is a condition of rest or motion with constant velocity (magnitude and direction). The vector sum of all forces acting on an object in equilibrium is zero.

$$\Sigma \mathbf{F} = 0 \tag{4.9}$$

In the case of two-dimensional equilibrium the sum of all forces in the x and y directions must separately equal zero.

$$\Sigma F_x = 0 \quad \text{and} \quad \Sigma F_y = 0 \tag{4.10}$$

The force of static friction between two surfaces in contact but not in motion, relative to each other, cannot be greater than $\mu_s n$, where n is the normal (perpendicular) force between the two surfaces and μ_s (coefficient of static friction) is a dimensionless constant which depends on the nature of the pair of surfaces.

$$f_s \leq \mu_s n \tag{4.11}$$

When two surfaces are in relative motion, the force of kinetic friction on each body is directed opposite to the direction of motion of the body. The coefficient of kinetic friction, μ_k, depends on the nature of the two surfaces.

$$f_k = \mu_k n \tag{4.12}$$

SUGGESTIONS, SKILLS, AND STRATEGIES

For problems involving **objects in equilibrium**.

1. Make a sketch of the situation described in the problem statement.

2. Draw a free-body diagram for the isolated object under consideration and label all external forces acting on the object. Try to guess the correct direction for each force. If you select a direction that leads to a negative sign in your solution for a force, do not be alarmed; this merely means that the direction of the force is the opposite of what you assumed.

3. Choose a convenient direction for a coordinate system and resolve all forces into x and y components.

4. Use the equations $\Sigma F_x = 0$ and $\Sigma F_y = 0$ (for objects in equilibrium, acceleration equals zero). Remember to keep track of the signs of the various force components.

5. Application of Step 4 above leads to a set of equations with several unknowns. All that is left is to solve the simultaneous equations for the unknowns in terms of the known quantities.

For problems involving the **application of Newton's second law**:

1. Draw a simple, neat diagram of the system.

2. Isolate the object of interest whose motion is being analyzed. Draw a free-body diagram for this object; that is, a diagram showing **all external forces acting on the object**. For systems containing more than one object, draw a **separate** diagram for each object. **Do not include forces which the object of interest exerts on other objects.**

3. Establish convenient coordinate axes for each object and find the components of the forces along these axes. It is usually convenient to choose the coordinate system so that one of the axes is parallel to the direction of motion of the object.

4. Apply Newton's second law, $\Sigma \mathbf{F} = m\mathbf{a}$, in the x and y directions for each object under consideration.

5. Solve the component equations for the unknowns. Remember that you must have as many independent equations as you have unknowns in order to obtain a complete solution.

6. Often in solving such problems, one must also use the equations of kinematics (motion with constant acceleration) from Chapter 2 to find all the unknowns.

REVIEW CHECKLIST

▷ State in your own words a description of Newton's laws of motion, recall physical examples of each law, and identify the action-reaction force pairs in a multiple-body interaction problem as specified by Newton's third law. (Sections 4.2, 4.3, and 4.4)

▷ Express the normal force in terms of other forces acting on an object and write out the equation which relates the coefficient of friction, force of friction and normal force between an object and the surface on which it rests or moves. (Section 4.6)

▷ Apply Newton's laws of motion to various mechanical systems using the recommended procedure discussed in Section 4.5. Most important, you should identify all external forces acting on the object of interest. Draw the **correct** free-body diagrams which apply to each body of the system, and apply Newton's second law, $\Sigma \mathbf{F} = m\mathbf{a}$, in **component** form. (Section 4.5)

▷ Apply the equations of kinematics (which involve the quantities displacement, velocity, and acceleration) as described in Chapter 2 along with those methods and equations of Chapter 4 (involving mass, force, and acceleration) to the solutions of problems where **both** the kinematic and dynamic aspects are present. (Sections 4.5 and 4.6)

▷ Be familiar with solving several linear equations simultaneously for the unknown quantities. Recall that you must have as many **independent** equations as you have unknowns.

SOLUTIONS TO SELECTED END-OF-CHAPTER PROBLEMS

5.   A bag of sugar weighs 5.00 lb on Earth. What should it weigh in newtons on the Moon, where the free-fall acceleration is 1/6 that on Earth? Repeat for Jupiter, where g is 2.64 times that on Earth. Find the mass of the bag of sugar in kilograms at each of the three locations.

Solution   To solve this problem, one must realize that the weight, w, of an object located somewhere in space is given by $w = mg$, where m is the mass of the object and g is the acceleration due to gravity at that location. The mass is a constant property of the object. The acceleration due to gravity, and therefore the weight of the object, varies as the object moves to different points in space.

If w_E is the weight of the object on Earth where the acceleration due to gravity is g_E and w_M is its weight on the moon where gravity is g_M, the ratio of these weights is

$$\left(\frac{w_M}{w_E}\right) = \left(\frac{mg_M}{mg_E}\right) = \frac{g_M}{g_E}$$

The weight on the moon is then given by $\qquad w_M = w_E\left(\frac{g_M}{g_E}\right) = \frac{w_E}{6}$

In an identical manner, the weight of the object on Jupiter will be:

$$w_J = w_E\left(\frac{g_J}{g_E}\right) = 2.64 w_E$$

Thus, a bag of sugar that has a weight on Earth of $w_E = 5.00 \text{ lbs} = 5.00 \text{ lbs}\left(\dfrac{4.448 \text{ N}}{1.00 \text{ lb}}\right)$

or $\qquad\qquad w_E = 22.24 \text{ N}$

will have a weight on the moon or on Jupiter, of $\qquad w_M = \dfrac{22.24 \text{ N}}{6} = 3.71 \text{ N}$

and $\qquad\qquad w_J = 2.64(22.24 \text{ N}) = 58.7 \text{ N} \qquad\qquad \Diamond$

The constant mass of the object may be found as: $\quad m = \dfrac{w_E}{g_E} = \dfrac{22.2 \text{ N}}{9.80 \text{ m}/\text{s}^2} = 2.27 \text{ kg} \qquad \Diamond$

---

11.   A boat moves through the water with two forces acting on it. One is a 2000-N forward push by the water on the propeller, and the other is an 1800-N resistive force due to the water around the bow. (a) What is the acceleration of the 1000-kg boat? (b) If it starts from rest, how far will it move in 10.0 s? (c) What will its velocity be at the end of this time?

Solution   A resistive force is always directed opposite to the motion. The forward push of the motor on the boat is in the direction of the motion, so the horizontal forces acting on the boat are as shown in the sketch at the right.

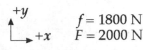

(a)   Taking the forward direction as positive, application of Newton's second law gives the horizontal acceleration as

$$a_x = \frac{\Sigma F_x}{m} = \frac{+2000\ \text{N} - 1800\ \text{N}}{1000\ \text{kg}} = +0.200\ \text{m / s}^2 \qquad \Diamond$$

(b)   The horizontal displacement of the boat during the first 10.0 s after it starts from rest is

$$\Delta x = v_{ix}t + \frac{1}{2}a_x t^2 = 0 + \frac{1}{2}(0.200\ \text{m / s}^2)(10.0\ \text{s})^2 = 10.0\ \text{m} \qquad \Diamond$$

(c)   At the end of this 10.0-s time interval, the horizontal speed of the boat is

$$v_x = v_{ix} + a_x t = 0 + (0.200\ \text{m / s}^2)(10.0\ \text{s}) = 2.00\ \text{m / s} \qquad \Diamond$$

---

17.   A 150 N bird feeder is supported by three cables as shown in Figure P4.17. Find the tension in each cable.

Solution   The bird feeder as well as the junction in the supporting cables are held in equilibrium by the forces acting on them. Thus, $\Sigma F_x = 0$ and $\Sigma F_y = 0$ may be applied to each of these objects. Consider the free-body diagrams (a) and (b) for these objects.

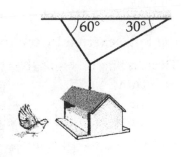

Figure P4.17

(a)   Free-body diagram of feeder   (b)   Free body diagram of junction

Note: Newton's third law has been observed in the directions of the action-reaction forces labeled T_1 in these diagrams.

Considering diagram (a) gives:

$$\sum F_y = +T_1 - 150 \text{ N} = 0 \qquad \text{or} \qquad T_1 = 150 \text{ N} \qquad \Diamond$$

Consideration of diagram (b) yields two equations:

$$\sum F_x = +T_2 \cos 30° - T_3 \cos 60° = 0 \qquad \sum F_y = T_2 \sin 30° + T_3 \sin 60° - T_1 = 0$$

which become, respectively,

$$T_3 = \left(\frac{\cos 30°}{\cos 60°}\right) T_2 = 1.73 \, T_2 \qquad \text{[1]} \qquad (0.500)T_2 + (0.866)T_3 = 150 \text{ N} \qquad \text{[2]}$$

Substituting [1] into [2], $\qquad\qquad [0.500 + (0.866)(1.73)]T_2 = 150 \text{ N}$

Thus, $\qquad\qquad\qquad\qquad\qquad\qquad T_2 = 75 \text{ N} \qquad \Diamond$

Then Equation [1] yields: $\qquad\qquad T_3 = (1.73)(75 \text{ N}) \qquad \text{or} \qquad T_3 = 130 \text{ N} \qquad \Diamond$

---

23. A 5.0-kg bucket of water is raised from a well by a rope. If the upward acceleration of the bucket is $3.0 \text{ m} / \text{s}^2$, find the force exerted by the rope on the bucket.

Solution

Choosing upward as positive, the acceleration of the bucket is

$$a_y = +3.0 \text{ m} / \text{s}^2$$

Newton's second law then gives the resultant vertical force acting on the bucket as

$$(F_R)_y = ma_y = (5.0 \text{ kg})(3.0 \text{ m} / \text{s}^2) = 15 \text{ N}$$

This resultant is also given by $(F_R)_y = \Sigma F_y = T - mg$

Thus, $\qquad\qquad\qquad\qquad T = (F_R)_y + mg = 15 \text{ N} + (5.0 \text{ kg})(9.80 \text{ m} / \text{s}^2) = 64 \text{ N} \qquad \Diamond$

33. A 1000-kg car is pulling a 300-kg trailer. Together the car and trailer have an acceleration of $2.15 \text{ m}/\text{s}^2$ in the forward direction. Neglecting frictional forces on the trailer, determine (a) the net force on the car; (b) the net force on the trailer; (c) the force exerted by the trailer on the car; (d) the resultant force exerted by the car on the road.

Solution

Choose the $+x$ direction to be horizontal and forward with the $+y$ direction upward. The acceleration of both the car and the trailer then has components of $a_x = +2.15 \text{ m}/\text{s}^2$ and $a_y = 0$.

(a) The net force on the car is in the direction of the car's acceleration (in the forward direction) and has the magnitude:

$$\left(F_{\text{car}}\right)_{\text{net}} = m_{\text{car}}a = (1000 \text{ kg})\left(2.15 \text{ m}/\text{s}^2\right) = 2.15 \times 10^3 \text{ N} \qquad \lozenge$$

(b) Likewise, the net force on the trailer is $\left(F_{\text{trailer}}\right)_{\text{net}} = m_{\text{trailer}}\, a$

   or $\left(F_{\text{trailer}}\right)_{\text{net}} = (300 \text{ kg})\left(2.15 \text{ m}/\text{s}^2\right) = 645 \text{ N}$ (also in forward direction) $\lozenge$

(c) Consider the free-body diagrams for the car and the trailer. The only horizontal force on the trailer is T, the tension in the link connecting the car and trailer.

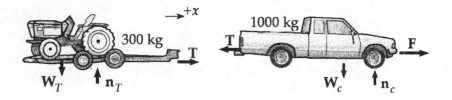

Thus, $T = \left(F_{\text{trailer}}\right)_{\text{net}} = 645 \text{ N}$ is the magnitude of the force exerted on the trailer by the car. By Newton's third law, the trailer exerts a force

   $T = 645 \text{ N}$ acting in the rearward direction on the car. $\lozenge$

(d) The road exerts the forward force F and the normal force n_c on the car. The magnitude of these forces may be found as follows:

$$\sum F_x = ma_x: \qquad F - 645 \text{ N} = (1000 \text{ kg})\left(+2.15 \text{ m}/\text{s}^2\right)$$

   or $\qquad\qquad F = 2.80 \times 10^3 \text{ N}$

$$\sum F_y = ma_y: \qquad +n_c - W_c = 0$$

so
$$n_c = m_{car}g = (1000 \text{ kg})(9.80 \text{ m / s}^2) = 9.80 \times 10^3 \text{ N}$$

The resultant force exerted on the car by the road is (by the vector diagram):

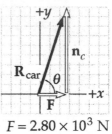

$$F = \sqrt{(2.80 \times 10^3 \text{ N})^2 + (9.80 \times 10^3 \text{ N})^2} = 1.02 \times 10^4 \text{ N}$$

at $\quad \theta = \tan^{-1}(n_c/F) = \tan^{-1}(3.50) = 74.1°$

$F = 2.80 \times 10^3$ N
$n_c = 9.80 \times 10^3$ N

By Newton's third law, the resultant force exerted on the road by the car is therefore 1.02×10^4 N directed at 74.1° below the negative x direction (or equivalently at 15.9° to the rear of vertical). ◊

---

37. A 1000-N crate is being pushed across a level floor at a constant speed by a force **F** of 300 N at an angle of 20.0° below the horizontal as shown in Figure P4.37a. (a) What is the coefficient of kinetic friction between the crate and the floor? (b) If the 300-N force is instead pulling the block at an angle of 20.0° above the horizontal as shown in Figure 4.37b, what will be the acceleration of the crate? Assume that the coefficient of friction is the same as found in (a).

Solution

(a) Figure (a) at the right is a free-body diagram of the crate in Figure P4.37a of the textbook. The crate is in equilibrium since it moves at constant velocity. Looking at the vertical forces, we can find the normal force:

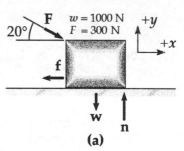

(a)

$$\sum F_y = +n - w - F\sin 20.0° = 0: \quad n = 1000 \text{ N} + (300 \text{ N})\sin 20.0° = 1.10 \times 10^3 \text{ N}$$

Then, considering the horizontal forces gives:

$$\sum F_x = +F\cos 20.0° - f = 0: \qquad f = (300 \text{ N})\cos 20.0° = 282 \text{ N}$$

Therefore, the coefficient of kinetic friction between the crate and floor is given by:

$$\mu_k = \frac{f}{n} = \frac{282 \text{ N}}{1.10 \times 10^3 \text{ N}} = 0.256 \qquad ◊$$

(b) If the 300-N force pulls upward at 20.0° above the horizontal, the free-body diagram is as given in figure (b). In this case, the vertical acceleration a_y is still zero, but the crate has some unknown horizontal acceleration. Considering the vertical forces,

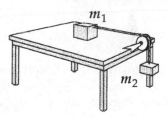

(b)

$$\sum F_y = ma_y = 0 \quad \text{giving} \qquad +F\sin 20.0° + n - w = 0$$

$$\text{or} \qquad n = 1000 \text{ N} - (300 \text{ N})\sin 20.0° = 897 \text{ N}$$

Then, assuming the same coefficient of friction as found in part (a), the friction force f is given by

$$f = \mu_k n = (0.256)(897 \text{ N}) = 230 \text{ N}$$

Noting that the mass of the crate is $m = w / g = (1000 \text{ N})/(9.80 \text{ m / s}^2) = 102 \text{ kg}$

apply Newton's second law to the horizontal motion of the crate: $\Sigma F_x = ma_x$

$$\text{or} \qquad a_x = \frac{F\cos 20.0° - f}{m} = \frac{(300 \text{ N})\cos 20.0° - (230 \text{ N})}{102 \text{ kg}} = 0.509 \text{ m / s}^2 \qquad \Diamond$$

---

45. Objects with masses $m_1 = 10.0$ kg and $m_2 = 5.00$ kg are connected by a light string that passes over a frictionless pulley as in Figure P4.30. If, when the system starts from rest, m_2 falls 1.00 m in 1.20 s, determine the coefficient of kinetic friction between m_1 and the table.

Solution

The free-body diagrams of the two objects in this system are shown at the right. Note that the accelerations of the two objects have the same magnitude, a, with the acceleration of m_1 directed horizontally to the right and the acceleration of m_2 directed vertically downward.

Since m_2 is observed to drop downward 1.00 m in 1.20 s when the system is released, the magnitude of the acceleration is found using $\Delta y = v_{iy}t + \frac{1}{2}a_y t^2$ as

$$-1.00 \text{ m} = 0 + \tfrac{1}{2}(-a)(1.20 \text{ s})^2 \qquad \text{so} \qquad a = 1.39 \text{ m / s}^2$$

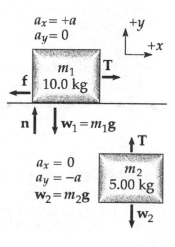

Figure P4.30

55

The weights of the objects are $w_1 = m_1 g = 98.0$ N and $w_2 = m_2 g = 49.0$ N. Applying Newton's second law to m_2 gives us the tension T in the cord.

$$\sum F_y = m_2 a_y: \quad +T - w_2 = m_2(-a)$$

$$T = 49.0 \text{ N} + (5.00 \text{ kg})(-1.39 \text{ m} / \text{s}^2) = 42.1 \text{ N}$$

Now, consider the vertical forces acting on m_1:

$$\sum F_y = m_1 a_y: \quad +n - w_1 = 0 \qquad \text{so} \qquad n = w_1 = 98.0 \text{ N}$$

Finally, considering the horizontal forces acting on m_1,

$$\sum F_x = m_1 a_x: \quad +T - f = m_1(+a)$$

$$f = T - m_1 a = 42.1 \text{ N} - (10.0 \text{ N})(1.39 \text{ m} / \text{s}^2)$$

so the friction force is $f = 28.2$ N. The coefficient of kinetic friction between m_1 and the table is therefore

$$\mu_k = \frac{f}{n} = \frac{28.2 \text{ N}}{98.0 \text{ N}} = 0.287 \qquad\qquad \Diamond$$

---

49. Find the acceleration experienced by each of the two objects shown in Figure P4.49 if the coefficient of kinetic friction between the 7.00-kg object and the plane is 0.250.

Solution

Free-body diagrams of each of the two masses in Figure P4.49 are given to the right. Here, the unknown magnitude of the accelerations of the masses is labeled a.

Note that it has been assumed that the 7.00-kg mass will accelerate up the incline and, consistent with that choice, it is assumed that the 12.0-kg mass will accelerate downward.

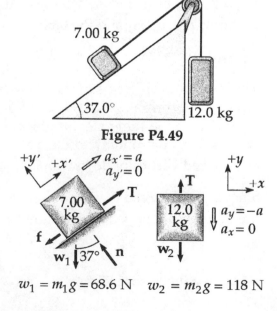

Figure P4.49

$w_1 = m_1 g = 68.6$ N $\quad w_2 = m_2 g = 118$ N

First, applying Newton's second law to the 7.00-kg mass,

$$\sum F_{y'} = m_1 a_{y'} \qquad \text{gives} \qquad +n - (68.6 \text{ N})\cos 37.0° = 0$$

or $$n = 54.8 \text{ N}$$

Therefore, the friction force f is: $\qquad f = \mu_k n = (0.250)(54.8 \text{ N}) = 13.7 \text{ N}$

Then, $\sum F_{x'} = m_1 a_{x'}$ gives $\qquad +T - f - (68.6 \text{ N})\sin 37.0° = m_1(+a)$

or $$T = 55.0 \text{ N} + (7.00 \text{ kg})a \qquad [1]$$

Next, applying Newton's second law to the 12.0-kg mass,

$$\sum F_y = m_2 a_y \qquad \text{yields} \qquad +T - 118 \text{ N} = (12.0 \text{ kg})(-a)$$

or $$T + (12.0 \text{ kg})a = 118 \text{ N} \qquad [2]$$

Substituting T in Equation [1] into Equation [2],

$$55.0 \text{ N} + (7.00 \text{ kg})a + (12 .0 \text{ kg})a = 118 \text{ N}$$

so $$(19.0 \text{ kg})a = 62.6 \text{ N}$$

Therefore, $$a = 3.30 \text{ m / s}^2$$

so the 7.00-kg mass accelerates up the incline at $\qquad 3.30 \text{ m / s}^2$

while the 12.0-kg mass accelerates downward at $\qquad 3.30 \text{ m / s}^2 \qquad \Diamond$

59. A box rests on the back of a truck. The coefficient of static friction between box and truck bed is 0.300. (a) When the truck accelerates forward, what force accelerates the box? (b) Find the maximum acceleration the truck can have before the box slides.

Solution

(a) Due to inertia, the box of mass m will tend to maintain its previous velocity (relative to the Earth) when the truck begins accelerating forward. Thus, the box will tend to slide toward the rear of the truck. The friction force exerted on the box by the truck bed will therefore be directed in the forward direction as it attempts to prevent this slippage.

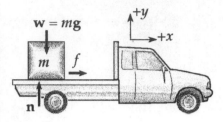

As seen in the free-body diagram of the box given in the sketch, this friction force f is the resultant horizontal force that will accelerate the box. ◊

(b) The box will have zero acceleration in the vertical (y) direction. Thus,

$$\sum F_y = n - w = 0 \qquad \text{gives} \qquad n = w = mg$$

Therefore, the maximum magnitude a static friction force between the box and truck bed can have is

$$f_{max} = \mu_s n = \mu_s mg$$

If the box has not started to slip, its horizontal acceleration a_x is the same as the acceleration, a, of the truck. Newton's second law,

$$\sum F_x = m a_x \qquad \text{then gives} \qquad f = ma \qquad \text{or} \qquad a = \frac{f}{m}$$

To find the maximum acceleration of the truck before slippage will occur, use the maximum static friction force, f_{max}, to obtain:

$$a_{max} = \frac{f_{max}}{m} = \frac{\mu_s mg}{m} = \mu_s g = (0.300)(9.80 \text{ m}/\text{s}^2) = 2.94 \text{ m}/\text{s}^2 \qquad ◊$$

---

63. A 3.00-kg block starts from rest at the top of a 30.0° incline and slides 2.00 m down the incline in 1.50 s. Find (a) the acceleration of the block, (b) the coefficient of kinetic friction between it and the incline, (c) the frictional force acting on the block, and (d) the speed of the block after it has slid 2.00 m.

Solution The sketch at the right shows the forces acting on the block. Notice that the positive x-direction has been chosen to be down the incline.

(a) Considering the 1.50-s interval after the block starts from rest,

$$\Delta x = v_{ix}t + \frac{1}{2}a_x t^2$$

$f_k = \mu_k n$

$w = mg = 29.4$ N

gives $a_x = \dfrac{2(\Delta x - v_{ix}t)}{t^2} = \dfrac{2(2.00\ \text{m} - 0)}{(1.50\ \text{s})^2} = 1.78\ \text{m}/\text{s}^2$ ◊

(c) Applying Newton's second law in the direction perpendicular to the incline,

$$\sum F_y = ma_y:$$ $n - mg\cos 30.0° = 0$

or $n = (3.00\ \text{kg})(9.80\ \text{m}/\text{s}^2)\cos 30.0° = 25.5$ N

Looking at the direction parallel to the incline,

$$\sum F_x = ma_x:$$ $mg\sin 30.0° - f_k = ma_x$

or $f_k = mg\sin 30.0° - ma_x$

$f_k = (29.4\ \text{N})\sin 30.0° - (3.00\ \text{kg})(1.78\ \text{m}/\text{s}^2) = 9.37$ N   ◊

(b) The coefficient of kinetic friction is then found to be

$$\mu_k = \frac{f_k}{n} = \frac{9.37\ \text{N}}{25.5\ \text{N}} = 0.368$$ ◊

(d) From $v_x^2 = v_{ix}^2 + 2a_x(\Delta x)$, the speed of the block after it has slid 2.00 m is:

$$v_x = \sqrt{0 + 2(1.78\ \text{m}/\text{s}^2)(2.00\ \text{m})} = 2.67\ \text{m}/\text{s}$$ ◊

---

74. On takeoff, the combined action of the air around the engines and wings of an airplane exerts an 8000-N force on the plane, directed upward at an angle of 65.0° above the horizontal. The plane rises with constant velocity in the vertical direction while continuing to accelerate in the horizontal direction. (a) What is the weight of the plane? (b) What is its horizontal acceleration?

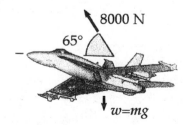

Solution The weight of the plane, $w = mg$, is the only force acting on the plane other than the 8000-N resultant force exerted by the engines and wings. Resolving the 8000-N force into its horizontal and vertical components gives

$$F_x = (8000 \text{ N}) \cos 65.0° = 3.38 \times 10^3 \text{ N} \qquad \text{and} \qquad F_y = (8000 \text{ N}) \sin 65.0° = 7.25 \times 10^3 \text{ N}$$

(a) Since the plane has constant velocity in the vertical direction (i.e., $a_y = 0$),

$$\sum F_y = m a_y \text{ yields } +F_y - w = 0 \qquad \text{or} \qquad w = F_y = 7.25 \times 10^3 \text{ N} \qquad \diamond$$

(b) The mass of the airplane is found as $\qquad m = \dfrac{w}{g} = \dfrac{7.25 \times 10^3 \text{ N}}{9.80 \text{ m} / \text{s}^2} = 740 \text{ kg}$

Then, applying Newton's second law to the horizontal motion gives

$$a_x = \frac{\sum F_x}{m} = \frac{3.38 \times 10^3 \text{ N}}{740 \text{ kg}} = 4.57 \text{ m} / \text{s}^2 \qquad \diamond$$

---

77. The board sandwiched between two other boards in Figure P4.77 weighs 95.5 N. If the coefficient of friction between the boards is 0.663, what must be the magnitude of the compression forces (assume horizontal) acting on both sides of the center board to keep it from slipping?

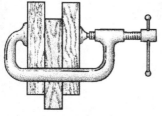

Solution Since the board is in equilibrium, $\Sigma F_x = 0$ and we see that the normal forces must be the same on both sides of the board. Also, if the minimum normal forces (compression forces) are being applied, the board is on the verge of slipping and the friction force on each side is

Figure P4.77

$$f = (f_s)_{max} = \mu_s n$$

The board is also in equilibrium in the vertical direction, so

$$\sum F_y = 2f - w = 0 \qquad \text{or} \qquad f = \frac{w}{2}$$

The minimum compression force needed is then

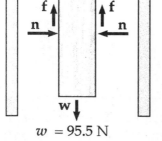

$w = 95.5 \text{ N}$

$$n = \frac{f}{\mu_s} = \frac{w}{2\mu_s} = \frac{95.5 \text{ N}}{2(0.663)} = 72.0 \text{ N} \qquad \diamond$$

Chapter 5
ENERGY

NOTES ON SELECTED CHAPTER SECTIONS

5.1   Work

In order for work to be accomplished, an object must undergo a displacement; the force associated with the work must have a component parallel to the direction of the displacement. Work is a scalar quantity and can be either positive or negative (positive when the component $F\cos\theta$ is in the same direction as the displacement). The SI unit of work is the newton-meter (N·m) or joule (J).

5.2   Kinetic Energy and the Work-Kinetic Energy Theorem

Any object which has mass m and speed v has kinetic energy. Kinetic energy is a scalar quantity and has the same units as work; kinetic energy of an object will change only if net work is done on the object by external forces. The relationship between work and change in kinetic energy is stated in the work-kinetic energy theorem.

5.3   Potential Energy

The work done on an object by the force of gravity is equal to the object's initial potential energy minus its final potential energy. The gravitational potential energy associated with an object depends only on the object's weight and its vertical height above the surface of the Earth. If the height above the surface increases, the potential energy will also increase; but the work done by the gravitational force will be negative. (In this case the direction of the displacement is opposite the direction of the gravitational force.) In working problems involving gravitational potential energy, it is necessary to **choose an arbitrary reference level (or location) at which the potential energy is taken to be zero**.

5.4   Conservative and Nonconservative Forces

A force is **conservative** if the work it does on an object moving between two points is independent of the path the object takes between the points. The work done on an object by a conservative force depends only on the initial and final positions of the object. The gravitational force is an example of a conservative force.

A force is **nonconservative** if the work it does on an object moving between two points depends on the path taken. Kinetic friction is an example of a nonconservative force.

5.5 Conservation of Mechanical Energy

The sum of the kinetic energy plus the potential energy is called the total mechanical energy.

The law of conservation of mechanical energy states that the mechanical energy of a system remains constant if the only forces that do work on the system are **conservative forces**.

5.6 Nonconservative Forces, Nonisolated Systems, and Conservation of Energy

In realistic situations, nonconservative forces such as friction are usually present, and the total mechanical energy of the system is not constant. In those cases, **the work done by all nonconservative forces equals the change in mechanical energy of the system, and Equation 5.15 must be used**; Equation 5.10 does not apply.

5.7 Power

Power delivered to an object is defined as the rate at which energy is transferred to the object or the rate at which work is being done on the object (see Equation 5.16). The **average power delivered to an object during a time interval** can be expressed as the product of the average speed during the time interval and the component of the force in the direction of the velocity (see Equation 5.17).

5.8 Work Done by a Varying Force

When work is done on an object by a force which is not constant, Equation 5.1 cannot be used to calculate the work done. If the value of the force is known as a function of the displacement, the work done by the varying force is the area under the Force-Displacement curve.

EQUATIONS AND CONCEPTS

The work done on a body by a force **F**, which is constant in both magnitude and direction, is defined to be the product of the component of the force in the direction of the displacement and the magnitude of the displacement.

$$W \equiv (F\cos\theta)\Delta x \qquad (5.1)$$

Sometimes a force is a function of position, as in the case for steel springs. In such a case, a plot of force vs. position may be made, and the work done by the force in moving from x_i to x_f is equal to the area under the curve.

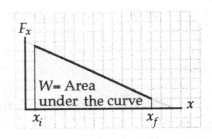

Note that the work done by a force can be positive, negative, or zero, depending on the value of θ, the angle between the direction of the force and the direction of the displacement. In the diagram at the right note the angle between the direction of the displacement and the force in each of the four cases as the object moves down the incline.

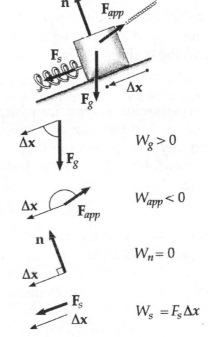

- If $0 \leq \theta < 90°$, W is positive.

- If $90° < \theta < 180°$, W is negative.

- If $\theta = 90°$ (**F** perpendicular to Δ**x**), $W = 0$.

- In the special case where **F** and Δ**x** are parallel and in the same direction, $\theta = 0$, $\cos\theta = 1$, and $W = F\Delta x$.

Work is a scalar quantity and the SI unit of work is the newton-meter or joule. See the summary of units of work in Table 5.1 of the textbook.

$$1\,\text{N} \cdot \text{m} = 1\,\text{J}$$

When a body of mass m experiences an acceleration due to a net force, the work done by the net force can be expressed in terms of the acceleration, mass, and the distance over which the acceleration is achieved.

$$W_{\text{net}} = F_{\text{net}}\Delta x = (ma)\Delta x \qquad (5.3)$$

The work done on a body by the net or resultant force can be expressed in terms of the change in the kinetic energy of the body. Kinetic energy is a scalar quantity and is the energy associated with an object's motion. Kinetic energy is defined by Equation 5.5. Equation 5.6 is a statement of the work-energy theorem, where it must be remembered that the work calculated by the equation is the work done by the net or resultant force acting on the body.

$$W_{net} = \frac{1}{2}mv^2 - \frac{1}{2}mv_0^2 \qquad (5.4)$$

$$KE \equiv \frac{1}{2}mv^2 \qquad (5.5)$$

$$W_{net} = KE_f - KE_i = \Delta KE \qquad (5.6)$$

A **conservative force** is one for which the work done by the force in moving between any two points is independent of the path followed from the initial to final point.

$$W_g = mgy_i - mgy_f \qquad (5.7)$$

Gravitational potential energy for a mass - Earth system near the surface of the Earth is defined by Equation 5.8. The force of gravity, $m\mathbf{g}$, is an example of a conservative force. The work done on a body by the force of gravity can be expressed in terms of initial and final values of the body's y-coordinates.

$$PE \equiv mgy \qquad (5.8)$$

$$W_g = PE_i - PE_f \qquad (5.9)$$

The units of energy (kinetic and potential) are the same as the units of work: $1\,J = 1\,N \cdot m$.

Comment on units.

In calculating the work done by the gravitational force, remember that the **difference in potential energy between two points is independent of the location of the origin**. Choose an origin which is convenient to calculate PE_i and PE_f for a particular situation.

Comment on reference level for potential energy.

When only conservative forces act on a system, the total mechanical energy ($KE+PE$) of the system remains constant; this is a statement of the law of conservation of mechanical energy.

$$KE_i + PE_i = KE_f + PE_f \qquad (5.10)$$

If the force of gravity (a conservative force) is the only force doing work within the system, the equation for conservation of mechanical energy takes a special form.

$$\tfrac{1}{2}mv_i^2 + mgy_i = \tfrac{1}{2}mv_f^2 + mgy_f \qquad (5.11)$$

The restoring force exerted by a stretched or compressed spring is proportional to the displacement of the "free" end of the spring and is directed opposite the displacement.

$$F_s = -kx \qquad (5.12)$$

The work done by a force in stretching or compressing a spring is stored in the spring as elastic potential energy. For a given displacement from the equilibrium position, the potential energy in the spring depends on the spring constant, k.

$$PE_s = \tfrac{1}{2}kx^2 \qquad (5.13)$$

If both conservative forces and nonconservative forces act on a system, the total mechanical energy will not remain constant. In this case, **the work done by all nonconservative forces equals the change in the total mechanical energy of the system.**

$$W_{nc} + W_c = \Delta KE$$

$$W_{nc} = (KE_f + PE_f) - (KE_i + PE_i) \qquad (5.15)$$

The average power supplied by a force is the ratio of the work done by the force to the time interval over which the force acts. The average power can also be expressed in terms of the force and the average speed of the object on which the force acts. In Equation 5.17, F is the component of the force along the direction of the velocity.

$$\overline{\mathcal{P}} \equiv \frac{W}{\Delta t} \tag{5.16}$$

$$\overline{\mathcal{P}} = F\overline{v} \tag{5.17}$$

The SI unit of power is the watt; in the British engineering system, the unit of power is the horsepower.

$$1\,W = 1\,J\,/\,s = 1\,kg \cdot m^2\,/\,s^3 \tag{5.18}$$

$$1\,hp = 550\,ft \cdot lb\,/\,s = 746\,W \tag{5.19}$$

SUGGESTIONS, SKILLS, AND STRATEGIES

CHOOSING A ZERO LEVEL

In working problems involving gravitational potential energy, it is always necessary to choose a location at which the gravitational potential energy is zero. This choice is completely arbitrary because the important quantity is the **difference** in potential energy, and that difference is independent of the location of zero. It is often convenient, but not essential, to choose the surface of the Earth as the reference position for zero potential energy. In most cases, the statement of the problem suggests a convenient level to use.

CONSERVATION OF ENERGY

Take the following steps in applying the principle of conservation of energy:

1. Define your system, which may consist of more than one object.

2. Select a reference position for the zero point of gravitational potential energy. This level must not be changed during the solution of a specific problem.

3. Determine whether or not nonconservative forces are present.

4. If mechanical energy is conserved (that is, if only conservative forces are present), you can write the total initial energy, $KE_i + PE_i$, at some point as the sum of the kinetic and potential energies at that point. Then, write an expression for the total final energy, $KE_f + PE_f$, at the final point of interest. Since mechanical energy is conserved, you can equate the two total energies and solve for the unknown.

5. If nonconservative forces such as friction are present (and thus mechanical energy is not conserved), first write expressions for the total initial and total final energies. In this case, the **difference between the two total energies is equal to the work done by the nonconservative force(s). That is, you should apply Equation 5.14.**

REVIEW CHECKLIST

▷   Define the work done by a constant force and work done by a force which varies with position. (Recognize that the work done by a force can be positive, negative, or zero; describe at least one example of each case.)

▷   Recognize that the gravitational potential energy of the mass-Earth system, $PE_g = mgy$, can be positive, negative, or zero, depending on the location of the reference level used to measure y. Be aware of the fact that although PE depends on the origin of the coordinate system, the **change** in potential energy, $(PE)_f - (PE)_i$, is **independent** of the coordinate system used to define PE.

▷   Understand that a force is said to be **conservative** if the work done by that force on a body moving between any two points is independent of the path taken. **Nonconservative** forces are those for which the work done on a particle moving between two points depends on the path. Account for nonconservative forces acting on a system using the work-kinetic energy theorem. In this case, the work done by all nonconservative forces equals the change in total mechanical energy of the system.

▷   Relate the work done by the net force on an object to the **change** in kinetic energy. The relation $W_{net} = \Delta KE = KE_f - KE_i$ is called the work-kinetic energy theorem, and is valid whether or not the (resultant) force is constant. That is, if we know the net work done on a particle as it undergoes a displacement, we also know the **change** in its kinetic energy. This is the most important concept in this chapter, so you must understand it thoroughly.

SOLUTIONS TO SELECTED END-OF-CHAPTER PROBLEMS

5. Starting from rest, a 5.00-kg block slides 2.50 m down a rough 30.0° incline. The coefficient of kinetic friction between the block and the incline is $\mu_k = 0.436$. Determine (a) the work done by the force of gravity, (b) the work done by the friction force between block and incline, and (c) the work done by the normal force.

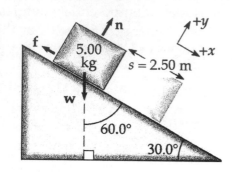

Solution

(a) The force of gravity is

$$w = mg = (5.00 \text{ kg})(9.80 \text{ m/s}^2) = 49.0 \text{ N}$$

(directed straight downward.)

The work done by this force is given by $W_g = ws\cos\theta$, where s is the displacement of the object and θ is the angle between the direction of the gravitational force and the direction of the displacement.

Thus,

$$W_g = (49.0 \text{ N})(2.50 \text{ m})\cos 60.0° = 61.3 \text{ J} \qquad \Diamond$$

(b) To find the friction force f, it is first necessary to solve for the normal force, n.

Using Newton's second law and recognizing that the block has zero acceleration directed perpendicular to the incline,

$$\sum F_y = n - w\sin 60.0° = 0: \qquad n = w\sin 60.0° = (49.0 \text{ N})\sin 60.0° = 42.4 \text{ N}$$

The friction force is then

$$f = \mu_k n = (0.436)(42.4 \text{ N}) = 18.5 \text{ N}$$

and the work done by it is

$$W_f = fs\cos\theta$$

where θ is now the angle between the directions of the friction force and the displacement.

Therefore,

$$W_f = (18.5 \text{ N})(2.50 \text{ m})\cos 180° = -46.3 \text{ J} \qquad \Diamond$$

(c) The normal force is perpendicular to the incline, and hence perpendicular to the displacement. The work done by the normal force is therefore

$$W_n = ns\cos 90° = 0 \qquad \Diamond$$

11.   A person doing a chin-up weighs 700 N exclusive of the arms. During the first 25.0 cm of the lift, each arm exerts an upward force of 355 N on the torso. If the upward movement starts from rest, what is the person's velocity at this point?

Solution

Three forces act on the torso of the person. These are the two 355 N forces, exerted upward by the arms, and a downward gravitational force (weight), $w = 700$ N. As the torso undergoes an upward displacement of $s = 0.250$ m, the **net** work done on it by these forces is

$$W_{net} = W_{arms} + W_{gravity} = 2F_{arm}\, s\cos\theta_1 + ws\cos\theta_2$$

The forces exerted by the arms are in the same direction as the displacement, so $\theta_1 = 0°$. The gravitational force is directed opposite to the displacement, or $\theta_2 = 180°$.

The net work done on the torso is then:

$$W_{net} = 2(355\ \text{N})(0.250\ \text{m})\cos 0° + (700\ \text{N})(0.250\ \text{m})\cos 180° = +2.50\ \text{J}$$

The mass of the torso is $m = \dfrac{w}{g}$:
$$m = \frac{700\ \text{N}}{9.80\ \text{m}/\text{s}^2} = 71.4\ \text{kg}$$

Applying the work-kinetic energy theorem,
$$W_{net} = KE_f - KE_i$$

gives: $W_{net} = \frac{1}{2}mv_f^2 - \frac{1}{2}mv_i^2$ or $2.50\ \text{J} = \frac{1}{2}(71.4\ \text{kg})v_f^2 - 0$

Therefore,
$$v_f^2 = \frac{2(2.50\ \text{J})}{71.4\ \text{kg}} = 0.070\ \text{m}^2/\text{s}^2$$

and
$$\mathbf{v}_f = 0.265\ \text{m}/\text{s}\ \textbf{upward}\qquad \Diamond$$

---

17.   A 2000-kg car moves down a level highway under the actions of two forces. One is a 1000-N forward force exerted on the drive wheels by the road; the other is a 950-N resistive force. Use the work-kinetic energy theorem to find the speed of the car after it has moved a distance of 20 m, assuming it starts from rest.

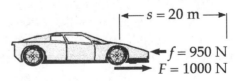

Solution   The forward force exerted on the car by the road is in the direction of the displacement $(\theta_1 = 0°)$ while the resistive force is directed opposite to the displacement $(\theta_2 = 180°)$. Other forces acting on the car (i.e., its weight and the normal force exerted by the road) are perpendicular to the displacement and, hence, do no work.

Therefore, the net work done on the car is

$$W_{net} = (F\cos\theta_1)s + (f\cos\theta_2)s: \qquad W_{net} = [(1000\text{ N})\cos 0°]s + [(950\text{ N})\cos 180°]s$$

$$W_{net} = (1000\text{ N} - 950\text{ N})(20\text{ m}) = 1.0 \times 10^3\text{ J}$$

The work-kinetic energy theorem then gives

$$W_{net} = \tfrac{1}{2}mv_f^2 - \tfrac{1}{2}mv_i^2 = \tfrac{1}{2}mv_f^2 - 0$$

or

$$v_f = \sqrt{\frac{2W_{net}}{m}} = \sqrt{\frac{2(1.0 \times 10^3\text{ J})}{2000\text{ kg}}} = 1.0\text{ m / s} \qquad \Diamond$$

---

24.   A softball pitcher rotates a 0.250-kg ball around a vertical circular path of radius 0.600 m before releasing it. The pitcher exerts a 30.0-N force directed parallel to the motion of the ball around the complete circular path. The speed of the ball at the top of the circle is 15.0 m/s. If the ball is released at the bottom of the circle, what is its speed upon release?

Solution   The speed of the ball at the point of release is most easily found using the work-kinetic energy theorem,

$$W_{net} = W_c + W_{nc} = KE_f - KE_i = \tfrac{1}{2}mv_f^2 - \tfrac{1}{2}mv_i^2$$

As the ball moves from the highest to the lowest point on the circular path, one conservative force (the weight of the ball, $w = mg = 2.45\text{ N}$) acts on it. The work done by this force is equal to the decrease in the gravitational potential energy,

$$W_c = PE_i - PE_f = mg(y_i - y_f) = w(2R)$$

The radius of the circular path is $R = 0.600\text{ m}$, so the work done by conservative forces is

$$W_c = (2.45\text{ N})(1.20\text{ m}) = 2.94\text{ J}$$

At any point along the circular path, the non-conservative force exerted by the pitcher's arm has a radial component (directed toward the center of the circular path) and a component that is tangential to the path. The radial component of this force, **C**, is always perpendicular to the motion and does no work. It is given that the tangential component has constant magnitude, $F = 30.0 \text{ N}$. This component is always parallel to the motion and acts on the ball through a distance equal to one-half the circumference of the circular path, $s = \pi R$. The work done by this non-conservative force is

$$W_{nc} = Fs = F(\pi R) = (30.0 \text{ N})\pi(0.600 \text{ m}) = 56.5 \text{ J}$$

The work-kinetic energy theorem then gives: $\qquad W_{net} = W_c + W_{nc} = \frac{1}{2}m\left(v_f^2 - v_i^2\right)$

or $\qquad v_f^2 = v_i^2 + \dfrac{2(W_c + W_{nc})}{m} = (15.0 \text{ m / s})^2 + \dfrac{2(2.94 \text{ J} + 56.5 \text{ J})}{0.250 \text{ kg}} = 700 \text{ m}^2/\text{ s}^2$

and $\qquad v_f = 26.5 \text{ m / s}$ $\qquad\qquad\qquad\qquad\qquad\qquad\qquad\qquad\qquad$ ◊

---

27. A child and sled with a combined mass of 50.0 kg slide down a frictionless slope. If the sled starts from rest and has a speed of 3.00 m/s at the bottom, what is the height of the hill?

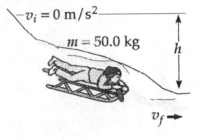

Solution As the child-sled combination slides down the frictionless hill, only one non-conservative force acts on it. This is the normal force exerted by the ground. It is always perpendicular to the displacement and hence does zero work:

$$(W = Fs\cos\theta = 0 \qquad \text{if} \qquad \theta = 90°)$$

Then, since no non-conservative forces do work on the system, the total mechanical energy is conserved:

$$KE_f + PE_f = KE_i + PE_i$$

Choose the initial state to be when the system starts from rest at the top of the hill and the final state to be when the system reaches the bottom. Also, choose $y = 0$ to be at the bottom of the hill. Then $v_i = 0$, $v_f = 3.00 \text{ m / s}$, $y_i = h$, and $y_f = 0$, where h is the height of the hill. The conservation of energy equation then reduces to:

$$\tfrac{1}{2}mv_f^2 + 0 = 0 + mgh \qquad \text{or} \qquad h = \frac{v_f^2}{2g} = \frac{(3.00 \text{ m / s})^2}{2(9.80 \text{ m / s}^2)} = 0.459 \text{ m} \qquad ◊$$

---

33. The launching mechanism of a toy gun consists of a spring of unknown spring constant, as shown in Figure P5.33a. If the spring is compressed a distance of 0.120 m and the gun fired vertically as shown, the gun can launch a 20.0-g projectile from rest to a maximum height of 20.0 m above the starting point of the projectile. Neglecting all resistive forces, determine (a) the spring constant and (b) the speed of the projectile as it moves through the equilibrium position of the spring (where $x = 0$), as shown in Figure P5.33b.

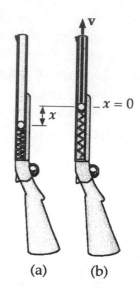

(a)   (b)

Figure P5.33

Solution Since resistance forces may be neglected, the mechanical energy of the system (gun and ball) will be conserved:

$$\left(KE + PE_g + PE_s\right)_{final} = \left(KE + PE_g + PE_s\right)_{initial}$$

(a) Choose the initial state to be the instant the ball starts from rest (with spring compressed) and the final state to be the instant the ball reaches maximum height (spring relaxed). With these choices, $v_i = v_f = 0$, $y_f - y_i = 20.0$ m, $x_i = 0.120$ m, and $x_f = 0$. The energy conservation equation then reduces to

$$0 + mgy_f + 0 = 0 + mgy_i + \frac{1}{2}kx_i^2$$

and gives the spring constant as

$$k = \frac{2mg\left(y_f - y_i\right)}{x_i^2} = \frac{2\left(20.0 \times 10^{-3} \text{ kg}\right)(20.0 \text{ m})}{(0.120 \text{ m})^2} = 544 \text{ N / m} \qquad \lozenge$$

(b) In this case, select the initial state to be as before but take the final state to be the instant the ball passes the $x = 0$ level. Then $y_f - y_i = 0.120$ m and energy conservation yields

$$\frac{1}{2}mv_f^2 + mgy_f + 0 = 0 + mgy_i + \frac{1}{2}kx_i^2 \qquad \text{or} \qquad v_f = \sqrt{\frac{kx_i^2}{m} - 2g\left(y_f - y_i\right)}$$

$$v_f = \sqrt{\frac{(544 \text{ N / m})(0.120 \text{ m})^2}{20.0 \times 10^{-3} \text{ kg}} - 2\left(9.80 \text{ m / s}^2\right)(0.120 \text{ m})} = 19.7 \text{ m / s} \qquad \lozenge$$

39. A 70-kg diver steps off a 10-m tower and drops, from rest, straight down into the water. If he comes to rest 5.0 m beneath the surface, determine the average resistance force exerted on him by the water.

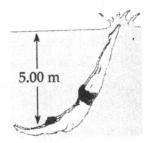

5.00 m

Solution

A non-conservative force acts on the diver during the 5.0-m movement through the water. This is the resistance force (directed upward) exerted by the water. The work done by this force on the diver is

$$W_{nc} = \overline{F}s\cos\theta = \overline{F}(5.0 \text{ m})\cos180° = -(5.0 \text{ m})\overline{F}$$

The only other force acting on the diver during his motion is his weight, which acts for the entire 15-m displacement. This is a conservative force, and the work done by it may be accounted for with potential energy terms in the work-kinetic energy theorem as follows:

$$W_{net} = W_{nc} + W_c = W_{nc} + \left(PE_i - PE_f\right) = KE_f - KE_i$$

Since the diver starts from rest and ends up at rest, $KE_f = KE_i = 0$

Setting $y = 0$ at the bottom of the dive, $PE_f = mgy_f = 0$

and the initial potential energy is

$$PE_i = mgy_i = (70 \text{ kg})(9.8 \text{ m} / \text{s}^2)(5.0 \text{ m} + 10 \text{ m}) = (690 \text{ N})(+15 \text{ m})$$

By the work-kinetic energy equation, $-(5.0 \text{ m})\overline{F} + (690 \text{ N})(15 \text{ m}) = 0$

and the average resistance force is $\overline{F} = (690 \text{ N})(15 \text{ m})/(5.0 \text{ m}) = 2.1 \times 10^3 \text{ N}$ ◊

45. A skier starts from rest at the top of a hill that is inclined at 10.5° with the horizontal. The hillside is 200 m long, and the coefficient of friction between snow and skis is 0.0750. At the bottom of the hill, the snow is level and the coefficient of friction is unchanged. How far does the skier move along the horizontal portion of the snow before coming to rest?

Solution

Select the reference level of gravitational potential energy ($PE_g = 0$) at the level of the base of the hill and let x represent the horizontal distance the skier travels after reaching this level.

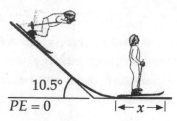

While on the hill, the normal force exerted on the skier by the snow is

$$n_1 = (mg)\cos 10.5°$$

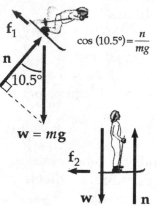

and the friction force is $\qquad f_1 = \mu_k n_1 = \mu_k (mg)\cos 10.5°$

On the level snow, the normal force is

$$n_2 = mg$$

and the friction force is $\qquad f_2 = \mu_k n_2 = \mu_k mg$

Consider the entire trip, from when the skier starts from rest on the hill until the skier comes to rest on the level snow.

Then, $\quad y_i - y_f = (200 \text{ m})\sin 10.5° \qquad$ and $\qquad\qquad v_i = v_f = 0$

The work done by non-conservative forces is

$$W_{nc} = f_1(200 \text{ m})\cos 180° + f_2(x)\cos 180° = -\mu_k(mg)\left[(200 \text{ m})\cos 10.5° + x\right]$$

Application of the work-kinetic energy theorem,

$$W_{nc} = \left(KE + PE_g\right)_f - \left(KE + PE_g\right)_i$$

then gives:

$$-\mu_k(mg)\left[(200 \text{ m})\cos 10.5° + x\right] = 0 + (mg)y_f - 0 - (mg)y_i = -(mg)(200 \text{ m})\sin 10.5°$$

or $\qquad x = (200 \text{ m})\left[\dfrac{\sin 10.5°}{\mu_k} - \cos 10.5°\right]$

With $\quad \mu_k = 0.0750 \qquad\qquad\qquad$ this yields $\qquad\qquad x = 289 \text{ m} \qquad\qquad\qquad \Diamond$

53. A 1.50×10^3-kg car starts from rest and accelerates uniformly to 18.0 m/s in 12.0 s. Assume that air resistance remains constant at 400 N during this time. Find (a) the average power developed by the engine and (b) the instantaneous power output of the engine at $t = 12.0$ s just before the car stops accelerating.

Solution

(a) The car's acceleration is
$$a = \frac{v_f - v_i}{t} = \frac{18.0 \text{ m/s} - 0}{12.0 \text{ s}} = 1.50 \text{ m/s}^2$$

The constant forward force being exerted by the engine is found from

$$\sum F_x = F_{engine} - F_{resistance} = m a_x \qquad \text{or} \qquad F_{engine} = F_{resistance} + ma$$

Thus, $F_{engine} = 400 \text{ N} + \left(1.50 \times 10^3 \text{ kg}\right)\left(1.50 \text{ m/s}^2\right) = 2.65 \times 10^3 \text{ N}$

The average velocity of the car is $\bar{v} = \frac{1}{2}\left(v_f + v_i\right) = \frac{1}{2}(18.0 \text{ m/s} + 0) = 9.00 \text{ m/s}$

so the average power output by the engine $\left(\bar{\mathcal{P}} = F_{engine}\bar{v}\right)$ is

$$\bar{\mathcal{P}} = \left(2.65 \times 10^3 \text{ N}\right)(9.00 \text{ m/s}) = \left(2.39 \times 10^4 \text{ W}\right)\left(\frac{1 \text{ hp}}{746 \text{ W}}\right) = 32.0 \text{ hp} \qquad \Diamond$$

(b) At $t = 12.0$ s, the instantaneous velocity of the car is $v = 18.0$ m/s and the instantaneous power output of the engine $\left(\mathcal{P} = F_{engine}v\right)$ is,

$$\mathcal{P} = \left(2.65 \times 10^3 \text{ N}\right)(18.0 \text{ m/s}) = \left(4.77 \times 10^4 \text{ W}\right)\left(\frac{1 \text{ hp}}{746 \text{ W}}\right) = 63.9 \text{ hp} \qquad \Diamond$$

---

55. The force acting on a particle varies as in Figure P5.55. Find the work done by the force as the particle moves (a) from $x = 0$ m to $x = 8.00$ m, (b) from $x = 8.00$ m to $x = 10.0$ m, and (c) from $x = 0$ m to $x = 10.0$ m.

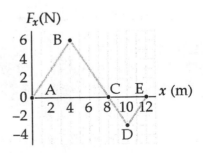

Figure P5.55

Solution

The work done on the particle by the force F as the particle moves from $x = x_i$ to $x = x_f$ is the area under the curve from x_i to x_f.

(a) For $\quad x = 0$ m $\qquad$ to $\qquad x = 8.00$ m

$$W = \text{area of triangle } ABC = \frac{1}{2}\overline{AC} \times \text{altitude}$$

$$W_{0 \to 8} = \frac{1}{2}(8.00 \text{ m})(6.00 \text{ N}) = 24.0 \text{ J}$$

◊

(b) For $\quad x = 8.00$ m $\qquad$ to $\qquad x = 10.0$ m

$$W_{8 \to 10} = \text{area of triangle } CDE = \frac{1}{2}\overline{CE} \times \text{altitude}$$

$$W_{8 \to 10} = \frac{1}{2}(2.00 \text{ m})(-3.00 \text{ N}) = -3.00 \text{ J}$$

◊

(c) For $\quad x = 0$ m $\qquad$ to $\qquad x = 10.0$ m

$$W_{0 \to 10} = W_{0 \to 8} + W_{8 \to 10} = 24.0 \text{ J} + (-3.00 \text{ J}) = 21.0 \text{ J}$$

◊

62. A toy gun uses a spring to project a 5.3-g soft rubber sphere horizontally. The spring constant is 8.0 N/m, the barrel of the gun is 15 cm long, and a constant frictional force of 0.032 N exists between barrel and projectile. With what speed does the projectile leave the barrel if the spring was compressed 5.0 cm for this launch?

Solution In this problem, only the spring force and the frictional force do work on the ball as the ball moves along the horizontal barrel. All other forces acting on the ball are perpendicular to the motion. The spring force is conservative and the work it does may be included in elastic potential terms in the work-kinetic energy theorem:

$$W_{net} = W_{nc} + W_c = W_{nc} + \left(PE_i - PE_f\right) = KE_f - KE_i$$

The elastic potential energy is $PE_s = \frac{1}{2}kx^2$ where k is the force constant of the spring and x is the amount the spring is stretched or compressed.

The frictional force is non-conservative, and the work it does is $W_{nc} = fs\cos180° = -fs$ since the friction force is directed opposite to the displacement. The ball is initially at rest, so $v_i = 0$ and $KE_i = \frac{1}{2}mv_i^2 = 0$. At the end, the spring is uncompressed so $x_f = 0$ and $\left(PE_s\right)_f = \frac{1}{2}kx_f^2 = 0$. Thus, the work-kinetic energy theorem gives:

$$-(0.032 \text{ N})(0.15 \text{ m}) + \left[\frac{1}{2}(8.0 \text{ N/m})(0.05 \text{ m})^2 - 0\right] = \frac{1}{2}\left(5.3 \times 10^{-3} \text{ kg}\right)v_f^2 - 0$$

or $\quad v_f^2 = 2.0 \text{ m}^2/\text{s}^2 \qquad$ and $\qquad v_f = 1.4 \text{ m/s}$

◊

70. A 5.0-kg block is pushed 3.0 m up a vertical wall with constant speed by a constant force of magnitude F applied at an angle of $\theta = 30°$ with the horizontal, as shown in Figure P5.70. If the coefficient of kinetic friction between block and wall is 0.30, determine the work done by (a) **F**, (b) the force of gravity, and (c) the normal force between block and wall. (d) By how much does the gravitational potential energy increase during this motion?

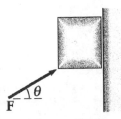

Figure P5.70

Solution

(a) Since the the block has zero acceleration,

$$\sum F_x = 0 \quad \text{so} \quad F\cos 30° - n = 0$$

$$\text{and} \quad n = 0.87F \quad \text{[1]}$$

Also, $\sum F_y = 0$ giving $F\sin 30° - w - f = 0$

$$\text{so} \quad 0.50F - f = w \quad \text{[2]}$$

But $w = mg = (5.0 \text{ kg})(9.8 \text{ m}/\text{s}^2) = 49 \text{ N}$

and $f = \mu_k n$ or, using Eq. [1], $f = 0.30(0.87F) = 0.26F$

Equation [2] then becomes $0.50F - 0.26F = 49 \text{ N}$

giving $F = 204 \text{ N}$

The work done by this force as the block moves 3.0 m up the wall is given by

$$W_F = Fs\cos\theta = (204 \text{ N})(3.0 \text{ m})\cos 60° = 3.1 \times 10^2 \text{ J} \qquad \Diamond$$

(b) The work done by the gravitational force (weight) as the block moves up the wall is

$$W_w = ws\cos\theta = (49 \text{ N})(3.0 \text{ m})\cos 180° = -1.5 \times 10^2 \text{ J} \qquad \Diamond$$

(c) Since the normal force n is perpendicular to the displacement, the work it does is

$$W_n = ns\cos\theta = ns\cos 90° = 0 \qquad \Diamond$$

(d) The change in the gravitational potential energy of the block is

$$\Delta PE = PE_f - PE_i = mgy_f - mgy_i = mg(\Delta y) = (49 \text{ N})(3.0 \text{ m}) = 1.5 \times 10^2 \text{ J} \qquad \Diamond$$

Note: The increase in the gravitational potential energy is the negative of the work done by the gravitational force (see part b).

77. In the dangerous "sport" of bungee-jumping, a daring student jumps from a balloon with a specially designed elastic cord attached to his waist, as shown in the figure. The unstretched length of the cord is 25.0 m, the student weighs 700 N, and the balloon is 36.0 m above the surface of a river below. Calculate the required force constant of the cord if the student is to stop safely 4.00 m above the river.

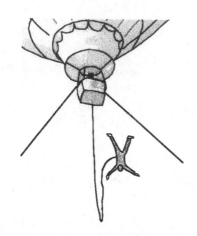

Solution Ignoring air resistance, there are no non-conservative forces acting on the student during the jump. Thus, the total mechanical energy is constant:

$$KE_f + \left(PE_g\right)_f + \left(PE_s\right)_f = KE_i + \left(PE_g\right)_i + \left(PE_s\right)_i$$

where

$$PE_g = \text{gravitational potential energy}$$

and

$$PE_s = \text{elastic potential energy}$$

The student is at rest both before and after the jump, so $\quad KE_f = KE_i = 0$

The cord is initially unstretched, (i.e., $x_i = 0$), $\qquad \left(PE_s\right)_i = \tfrac{1}{2}kx_i^2 = 0$

The conservation of energy equation then reduces to: $\qquad \left(PE_s\right)_f = \left(PE_g\right)_i - \left(PE_g\right)_f$

$$\text{or} \qquad \tfrac{1}{2}kx_f^2 = mgy_i - mgy_f$$

This may be rewritten as $\qquad \tfrac{1}{2}kx_f^2 = mg\left(y_i - y_f\right) = w\left(y_i - y_f\right)$

The student is to drop from 36.0 m down to 4.00 m above the river, so $y_i - y_f = 32.0 \text{ m}$. Also, the cord is 25.0 m long. Thus, when the student has dropped 32.0 m, the cord is stretched by

$$x_f = (32.0 - 25.0) \text{ m} = 7.00 \text{ m}$$

The energy equation then gives: $\qquad k = \dfrac{2w\left(y_i - y_f\right)}{x_f^2} = \dfrac{2(700 \text{ N})(32.0 \text{ m})}{(7.00 \text{ m})^2} = 914 \text{ N / m} \quad \Diamond$

Chapter 6
MOMENTUM AND COLLISIONS

NOTES ON SELECTED CHAPTER SECTIONS

6.1   Momentum and Impulse

The **time rate of change of the momentum** of a particle is equal to the **resultant force** on the particle. The **impulse** of a force is a vector quantity and is equal to the change in momentum of the particle on which the force acts. The impulse or change in momentum of an object is equal to the area under the force-time graph from the beginning to the end of the time interval during which the force is in contact with the object. Under the **impulse approximation**, it is assumed that one of the forces acting on a particle is of short time duration but of much greater magnitude than any of the other forces.

6.2   Conservation of Momentum
6.3   Collisions

The principle of **conservation of linear momentum** can be stated for an isolated system of objects. This is a system on which **no external forces** (e.g. friction or gravity) are acting. When no external forces act on a system the total linear momentum of the system remains constant. Remember, **momentum is a vector quantity** and the momentum of each individual particle may change but the total momentum of the entire system of particles remains constant. A collision between two or more masses is an important example of conservation of momentum.

For **any type of collision**, the total momentum before the collision equals the total momentum just after the collision.

In an **inelastic collision**, the total momentum is conserved; however, the total kinetic energy is not conserved.

In a **perfectly inelastic collision**, the two colliding objects stick together following the collision. This corresponds to a maximum loss in kinetic energy.

In an **elastic collision**, both momentum and kinetic energy are conserved.

6.4   Glancing Collisions

The law of conservation of momentum is not restricted to one-dimensional collisions. If two masses undergo a **two-dimensional** (glancing) **collision** and there are no external forces acting, the total momentum is conserved in each of the x, y, and z directions independently.

EQUATIONS AND CONCEPTS

An object of mass m and velocity $\mathbf{v}$ is characterized by a vector quantity called linear momentum. The SI units of linear momentum are kg·m/s.

$$\mathbf{p} \equiv m\mathbf{v} \qquad (6.1)$$

Since momentum is a vector quantity, the defining equation can be written in component form.

$$p_x = mv_x$$

$$p_y = mv_y$$

The resultant force acting on an object equals the time rate of change of the object's momentum. This equation is a mathematical expression of Newton's second law.

$$\mathbf{F}_{net} = \frac{\text{change in momentum}}{\text{time interval}} = \frac{\Delta \mathbf{p}}{\Delta t} \quad (6.2)$$

Note that as a special case in Equation 6.2, if the resultant force $\Sigma \mathbf{F} = 0$, then the momentum does not change.

Comment

Equation 6.4 is a mathematical statement of the important impulse-momentum theorem. The product of the net force acting on an object and the time interval during which it acts is called the **impulse** imparted to the object by the force and is equal to the **change in momentum** experienced by the object. Equations 6.4 and 6.5 are written for the case when a **single constant force** acts on an object.

$$\mathbf{F}\Delta t = \Delta \mathbf{p} = m\mathbf{v}_f - m\mathbf{v}_i \qquad (6.4)$$

$$\text{Impulse} = \mathbf{F}\Delta t \qquad (6.5)$$

The impulse imparted by a force during a time interval Δt is equal to the area under the force-time graph from the beginning to the end of the time interval. When the force varies in time, as illustrated in the figure, it is often convenient to define an average force, $\overline{F}$, which is a constant force that imparts the same impulse in a time interval Δt as the actual time varying force.

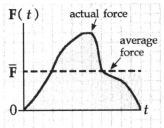

Comment on impulse of force

When two objects interact in a collision (and exert forces on each other) and **no external forces** act on the two-object system, the total momentum of the system before the collision equals the total momentum after the collision. This equation expresses the **principle of conservation of momentum**.

$$m_1 \mathbf{v}_{1i} + m_2 \mathbf{v}_{2i} = m_1 \mathbf{v}_{1f} + m_2 \mathbf{v}_{2f} \qquad (6.7)$$

In general, when the net external force acting on any system of objects is zero, the linear momentum of the system is conserved.

Comment on the law of conservation of momentum

An **elastic collision** is one in which both momentum and kinetic energy are conserved.

Comment on types of collisions.

An **inelastic collision** is one in which momentum is conserved but kinetic energy is not.

A **perfectly inelastic collision** is a collision in which the colliding objects stick together so that they have a common final velocity and the momentum of the system of objects is conserved.

Conservation of momentum for a one-dimensional **perfectly inelastic collision**.

$$m_1 v_{1i} + m_2 v_{2i} = (m_1 + m_2) v_f \qquad (6.8)$$

Note that v_{1i}, v_{2i}, and v_f are actually components of velocity vectors and hence they may have positive or negative values determined by the direction of motion relative to the chosen coordinate origin.

Comment on signs.

In an **elastic head-on collision**, both momentum and kinetic energy are conserved.

$$m_1 v_{1i} + m_2 v_{2i} = m_1 v_{1f} + m_2 v_{2f} \qquad (6.10)$$

$$\tfrac{1}{2} m_1 v_{1i}^2 + \tfrac{1}{2} m_2 v_{2i}^2 =$$
$$\tfrac{1}{2} m_1 v_{1f}^2 + \tfrac{1}{2} m_2 v_{2f}^2 \qquad (6.11)$$

This equation may be written to demonstrate another characteristic of a perfectly elastic collision: the relative velocity of the two objects before the collision equals the negative of the **relative velocity of the two objects** after the collision.

$$v_{1i} - v_{2i} = -(v_{1f} - v_{2f}) \qquad (6.14)$$

Consider a two-dimensional collision in which an object m_1 moves along the x axis and collides **perfectly elastically** with m_2 initially at rest. Momentum is conserved **along each direction**. Angles are defined in the diagram at the right.

x-component: $\qquad\qquad\qquad (6.15)$
$$m_1 v_{1i} + 0 = m_1 v_{1f} \cos\theta + m_2 v_{2f} \cos\phi$$

y-component: $\qquad\qquad\qquad (6.16)$
$$0 + 0 = m_1 v_{1f} \sin\theta - m_2 v_{2f} \sin\phi$$

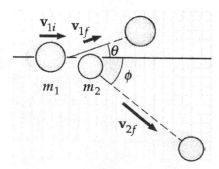

For a **perfectly elastic collision** kinetic energy is also conserved.

$$\frac{1}{2}m_1v_{1i}^2 = \frac{1}{2}m_1v_{1f}^2 + \frac{1}{2}m_1v_{2f}^2 \qquad (6.17)$$

SUGGESTIONS, SKILLS, AND STRATEGIES

The following procedure is recommended when dealing with problems involving collisions between two objects:

1. Set up a coordinate system and define velocities with respect to that system. That is, objects moving in the direction selected as the positive direction of the x axis are considered as having a positive velocity and negative if moving in the negative x direction. It is convenient to have the x axis coincide with the direction of one of the initial velocities.

2. In your sketch of the coordinate system, draw all velocity vectors with labels and include all the given information including scattering angles.

3. Write expressions for the momentum of each object before and after the collision. (In two-dimensional collision problems, write expressions for the x and y components of momentum before and after the collision.) Remember to include the appropriate signs for their velocity directions.

4. Now write expressions for the **total** momentum **before** and the total momentum **after** the collision and equate the two. (For two-dimensional collisions, these expressions should be written for the momentum in both the x and y directions. See Equations 6.15 and 6.16. It is important to emphasize that it is the momentum of the **system** (the two colliding objects) that is conserved, not the momentum of the individual objects.

5. If the collision is **perfectly inelastic** (kinetic energy is not conserved and the two objects have a common velocity following the collision) you should then proceed to solve the momentum equations for the unknown quantities.

6. If the collision is **elastic**, kinetic energy is also conserved, so you can equate the total kinetic energy before the collision to the total kinetic energy after the collision. This gives an additional relationship between the various velocities. The conservation of kinetic energy for elastic collisions leads to the expression $v_{1i} - v_{2i} = -(v_{1f} - v_{2f})$, which is often easier to use in solving elastic collision problems than is an expression for conservation of kinetic energy.

REVIEW CHECKLIST

▷ The impulse of a force acting on a particle during some time interval equals the **change** in momentum of the particle, and the impulse equals the area under the Force-Time graph.

▷ The momentum of any isolated system (one for which the net external force is zero) is conserved, regardless of the nature of the forces between the masses which comprise the system.

▷ There are two types of collisions that can occur between particles, namely elastic and inelastic collisions. Recognize that a **perfectly** inelastic collision is an inelastic collision in which the colliding particles stick together after the collision, and hence move as a composite particle. Kinetic energy is conserved in perfectly elastic collisions, but not in the case of inelastic collisions. **Linear momentum is conserved in both types of collisions, if the net external force acting on the system of colliding objects is zero.**

▷ The conservation of linear momentum applies not only to head-on collisions (one-dimensional), but also to glancing collisions (two- or three-dimensional). For example, in a two-dimensional collision, the total momentum in the x-direction is conserved and the total momentum in the y-direction is conserved.

▷ The equations for conservation of momentum and kinetic energy can be used to calculate the final velocities in a two-body head-on elastic collision; and to calculate the final velocity and the change of kinetic energy in a two-body system for a completely inelastic collision.

SOLUTIONS TO SELECTED END-OF-CHAPTER PROBLEMS

7. A professional diver performs a dive from a platform 10 m above the water surface. Estimate the order of magnitude of the average impact force she experiences in her collision with the water. State the quantities you take as data and their values.

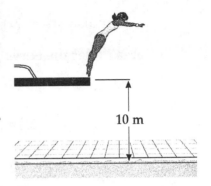

10 m

Solution The speed just before impact with the water is found from conservation of mechanical energy,

$$\left(KE + PE_g\right)_f = \left(KE + PE_g\right)_i$$

If the diver starts from rest at height h above the water, this yields

$$\tfrac{1}{2}mv^2_{impact} + 0 = 0 + mgh \qquad \text{or} \qquad v_{impact} = \sqrt{2gh}$$

As the diver is being brought to rest following impact with the water, she experiences an impulse given by

$$\overline{\mathbf{F}}(\Delta t) = \Delta \mathbf{p} = m\mathbf{v}_f - m\mathbf{v}_i = 0 - m\mathbf{v}_{impact}$$

Taking upward as the positive direction, the average force experienced by the diver during the interval, of duration Δt, between impact and the instant she comes to rest is

$$\overline{\mathbf{F}} = \frac{-m\mathbf{v}_{impact}}{\Delta t} = \frac{-m\left(-\sqrt{2gh}\right)}{\Delta t} = \frac{m\sqrt{2gh}}{\Delta t} \quad \text{(upward)}$$

Assuming a mass of 55 kg and an impact time of $\Delta t \cong 1.0$ s, the magnitude of this average force is

$$\overline{F} = \frac{(55\ \text{kg})\sqrt{2\left(9.80\ \text{m / s}^2\right)(10\ \text{m})}}{1.0\ \text{s}} = 770\ \text{N} \qquad \text{or} \qquad \overline{F} \sim 10^3\ \text{N} \qquad \Diamond$$

13. The forces shown in the force-time diagram in Figure P6.13 act on a 1.5-kg particle. Find (a) the impulse for the interval $t = 0$ to $t = 3.0$ s and (b) the impulse for the interval $t = 0$ to $t = 5.0$ s. (c) If the forces act on a 1.5-kg particle that is initially at rest, find the particle's speed at $t = 3.0$ s and at $t = 5.0$ s.

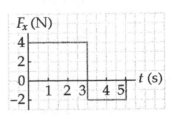

Figure P6.13

Solution

(a) The impulse is the area under the curve between $t = 0$ and $t = 3.0$ s. That is,

$$\text{Impulse} = (4.0 \text{ N})(3.0 \text{ s}) = 12 \text{ N} \cdot \text{s}$$ ◊

(b) The area under the curve between $t = 0$ and $t = 5.0$ s is:

$$\text{Impulse} = (4.0 \text{ N})(3.0 \text{ s}) + (-2.0 \text{ N})(2.0 \text{ s}) = 8.0 \text{ N} \cdot \text{s}$$ ◊

(c)

$$\text{Impulse} = \overline{F}(\Delta t) = \Delta p = m\left(v_f - v_i\right)$$

so $v_f = v_i + \dfrac{\text{Impulse}}{m}$ at 3.0 s: $v_f = v_i + \dfrac{\text{Impulse}}{m} = 0 + \dfrac{12 \text{ N} \cdot \text{s}}{1.5 \text{ kg}} = 8.0 \text{ m} / \text{s}$ ◊

at 5.0 s: $v_f = v_i + \dfrac{\text{Impulse}}{m} = 0 + \dfrac{8.0 \text{ N} \cdot \text{s}}{1.5 \text{ kg}} = 5.3 \text{ m} / \text{s}$ ◊

---

16. A pitcher throws a 0.15-kg baseball so that it crosses home-plate horizontally with a speed of 20 m/s. It is hit straight back at the pitcher with a final speed of 22 m/s. (a) What is the impulse delivered to the ball? (b) Find the average force exerted by the bat on the ball if the two are in contact for 2.0×10^{-3} s.

Solution

(a) **Choosing toward the batter as the positive direction,** the initial and final momenta of the ball are:

$$p_i = mv_i = (0.15 \text{ kg})(+20 \text{ m} / \text{s}) = +3.0 \text{ kg} \cdot \text{m} / \text{s}$$

and

$$p_f = mv_f = (0.15 \text{ kg})(-22 \text{ m} / \text{s}) = -3.3 \text{ kg} \cdot \text{m} / \text{s}$$

Then, the impulse delivered to the ball is

$$\text{Impulse} = p_f - p_i = (-3.3 \text{ kg} \cdot \text{m} / \text{s}) - (+3.0 \text{ kg} \cdot \text{m} / \text{s}) = -6.3 \text{ kg} \cdot \text{m} / \text{s}$$

or $\text{Impulse} = 6.3 \text{ kg} \cdot \text{m} / \text{s}$ directed **toward the pitcher.** ◊

(b) The impulse imparted to the ball may also be expressed as $\text{Impulse} = \overline{F}(\Delta t)$, where $\overline{F}$ is the average force acting on the ball and Δt is the duration of that force.

Thus, the average force exerted on the ball is

$$\overline{\mathbf{F}} = \frac{\text{Impulse}}{\Delta t} = \frac{-6.3 \text{ kg} \cdot \text{m} / \text{s}}{2.0 \times 10^{-3} \text{ s}} = -3.2 \times 10^3 \text{ N}$$

or $\qquad \overline{\mathbf{F}} = 3.2 \times 10^3 \text{ N}$ directed **toward the pitcher**. ◊

21.   A 45.0-kg girl is standing on a 150-kg plank. The plank, originally at rest, is free to slide on a frozen lake, which is a flat, frictionless surface. The girl begins to walk along the plank at a constant velocity to the right of 1.50 m/s relative to the plank. (a) What is her velocity relative to the ice surface? (b) What is the velocity of the plank relative to the ice surface?

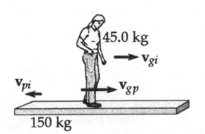

45.0 kg

150 kg

Solution

(a)   The velocity of the girl relative to the ice, v_{gi}, is $v_{gi} = v_{gp} + v_{pi}$ where v_{gp} is the velocity of the girl relative to the plank, and v_{pi} is the velocity of the plank relative to the ice.

Since we are given that $\qquad v_{gp} = 1.50 \text{ m} / \text{s}$

this becomes $\qquad v_{gi} = 1.50 \text{ m} / \text{s} + v_{pi}$ [1]

Conservation of momentum gives $\quad m_g v_{gi} + m_p v_{pi} = 0$

or $\qquad v_{pi} = -\left(\frac{m_g}{m_p} \right) v_{gi}$ [2]

Equation [1] becomes $\qquad \left(1 + \frac{m_g}{m_p} \right) v_{gi} = 1.50 \text{ m} / \text{s}$

or $\qquad v_{gi} = \frac{1.50 \text{ m} / \text{s}}{1 + \left(\dfrac{45.0 \text{ kg}}{150 \text{ kg}} \right)} = 1.15 \text{ m} / \text{s}$ ◊

(b)   Then, using Equation [2], $\qquad v_{pi} = -\left(\frac{45.0 \text{ kg}}{150 \text{ kg}} \right) (1.15 \text{ m} / \text{s}) = -0.346 \text{ m} / \text{s}$

or $\qquad v_{pi} = 0.346 \text{ m} / \text{s}$ (directed opposite the girl's motion) ◊

29. A 0.030-kg bullet is fired vertically at 200 m/s into a 0.15-kg baseball that is initially at rest. How high does the combination rise after the collision, assuming the bullet embeds itself in the ball?

Solution A common student error in a problem of this type is to attempt using conservation of **energy** from before the collision to the end of the motion. Since this is an inelastic collision, use of energy methods over any time interval that includes the collision is extremely difficult.

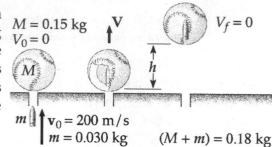

$M = 0.15$ kg $V_0 = 0$ $V_f = 0$

m $v_0 = 200$ m/s $m = 0.030$ kg $(M + m) = 0.18$ kg

The best approach is to **apply conservation of momentum from just before to just after the collision**. Then energy methods can be applied **after** the collision is past, if that is desirable.

Consider the sketches above: the velocity of the ball and embedded bullet, immediately after impact, is V. Using momentum conservation from just before impact to just after impact,

$$mv_0 + MV_0 = (M + m)V$$

Substituting in known values,

$$(0.030 \text{ kg})(200 \text{ m/s}) + 0 = (0.18 \text{ kg})V$$

which yields

$$V = \frac{6.0 \text{ kg} \cdot \text{m/s}}{0.18 \text{ kg}} = 33.3 \text{ m/s}$$

From just **after** the collision until the ball reaches the maximum height, only conservative forces do work on the ball-bullet combination.

Thus,

$$KE_f + PE_f = KE_i + PE_i$$

Choosing $y = 0$ at the initial level of the ball gives $y_i = 0$ and $y_f = h$;

so

$$PE_i = 0 \quad \text{and} \quad PE_f = (M + m)gh$$

Also,

$$V_f = 0 \quad \text{yields} \quad KE_f = 0$$

The energy equation then reduces to

$$0 + (M + m)gh = KE_i = \tfrac{1}{2}(M + m)V^2 + 0$$

or

$$h = \frac{V^2}{2g} = \frac{(33.3 \text{ m/s})^2}{2(9.8 \text{ m/s}^2)} = 57 \text{ m} \qquad \Diamond$$

37. A 25.0-g object moving to the right at 20.0 cm/s overtakes and collides elastically with a 10.0-g object moving in the same direction at 15.0 cm/s. Find the velocity of each object after the collision.

Solution The sketches to the right show the conditions just before and just after the collision. Note the choice of the $+x$ direction.

Applying conservation of momentum to this collision gives $\mathbf{p}_{1f} + \mathbf{p}_{2f} = \mathbf{p}_{1i} + \mathbf{p}_2$:

$$(25.0 \text{ g})\mathbf{v}_{1f} + (10.0 \text{ g})\mathbf{v}_{2f} = (25.0 \text{ g})(+20.0 \text{ cm/s}) + (10.0 \text{ g})(+15.0 \text{ cm/s})$$

which reduces to: $2.50\mathbf{v}_{1f} + \mathbf{v}_{2f} = +65.0 \text{ cm/s}$ **[1]**

This is a perfectly elastic, head-on collision,

so $\mathbf{v}_{1i} - \mathbf{v}_{2i} = -\left(\mathbf{v}_{1f} - \mathbf{v}_{2f}\right)$

or $+20.0 \text{ cm/s} - (+15.0 \text{ cm/s}) = -\mathbf{v}_{1f} + \mathbf{v}_{2f}$

This reduces to : $\mathbf{v}_{2f} = \mathbf{v}_{1f} + 5.00 \text{ cm/s}$ **[2]**

Substitution of Equation [2] into Equation [1] yields $3.50\mathbf{v}_{1f} = +60.0 \text{ cm/s}$

Thus, $\mathbf{v}_{1f} = +17.1 \text{ cm/s} = 17.1 \text{ cm/s in the } +x \text{ direction}$ ◊

Then, Equation [2] gives: $\mathbf{v}_{2f} = +22.1 \text{ cm/s} = 22.1 \text{ cm/s in } +x \text{ direction}$ ◊

---

43. A 2000-kg car moving east at 10.0 m/s collides with a 3000-kg car moving north. The cars stick together and move as a unit after the collision, at an angle of 40.0° north of east and at a speed of 5.22 m/s. Find the speed of the 3000-kg car before the collision.

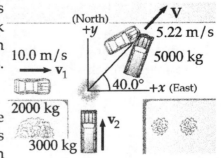

Solution The sketch at the right shows the conditions existing just before and just after this collision. Total momentum is conserved even in perfectly inelastic collisions.

Thus, $$\sum \mathbf{p}_f = \sum \mathbf{p}_i$$

This vector equation is equivalent to two component equations:

$$\left(\sum p_x\right)_f = \left(\sum p_x\right)_i \quad \text{and} \quad \left(\sum p_y\right)_f = \left(\sum p_y\right)_i$$

The only unknown in this problem is v_2, the magnitude of the initial velocity of the 3000-kg car. The direction of this velocity is known to be in the $+y$ (or north) direction. Thus, consider the equation for the momentum in the y-direction, which gives:

$$0 + (3000 \text{ kg})v_2 = (5000 \text{ kg})(5.22 \text{ m / s})\sin 40.0°$$

or $$v_2 = \frac{(5000 \text{ kg})(5.22 \text{ m / s})\sin 40.0°}{3000 \text{ kg}}$$

Solving, $v_2 = 5.59 \text{ m / s}$ ◊

Observe that knowledge of the initial speed of the 2000-kg car was unnecessary for this solution.

———————————————

45. A billiard ball moving at 5.00 m/s strikes a stationary ball of the same mass. After the collision, the first ball moves at 4.33 m/s at an angle of 30° with respect to the original line of motion. (a) Find the velocity (magnitude and direction) of the second ball after collision. (b) Was this an inelastic collision or an elastic collision?

Solution The sketch given to the right shows this glancing collision both just before impact and just after impact. The resultant momentum vector after collision is the same as the resultant momentum vector before collision, so

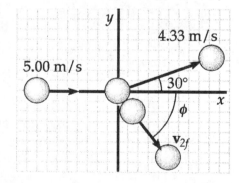

$$\left(\sum p_x\right)_f = \left(\sum p_x\right)_i \quad \text{and} \quad \left(\sum p_y\right)_f = \left(\sum p_y\right)_i$$

(a) First, looking at the x-components of momentum, this gives

$$m(4.33 \text{ m / s})\cos 30.0° + mv_{2f}\cos \phi = m(5.00 \text{ m / s}) + 0$$

Thus $$3.75 \text{ m / s} + v_{2f}\cos \phi = 5.00 \text{ m / s}$$

and $$v_{2f}\cos \phi = 1.25 \text{ m / s} \qquad [1]$$

Then, considering the y-components of momentum gives

$$m(4.33 \text{ m / s})\sin 30.0° + mv_{2f}\sin\phi = 0 + 0$$

which reduces to $v_{2f}\sin\phi = -2.17 \text{ m / s}$ [2]

Squaring Equations [1] and [2] and adding the results yields

$$v_{2f}^2\left(\cos^2\phi + \sin^2\phi\right) = (1.25 \text{ m / s})^2 + (-2.17 \text{ m / s})^2$$

Since (by the trigonometric identity) $\cos^2\phi + \sin^2\phi = 1$, the speed of the second ball after the collision is found as:

$$v_{2f}^2 = 6.25 \text{ m}^2/\text{s}^2 \quad \text{and} \quad v_{2f} = 2.50 \text{ m / s} \qquad \Diamond$$

Since we were asked for the velocity, we also need the angle ϕ of the velocity.

Dividing Equation [2] by Equation [1] gives $\dfrac{\sin\phi}{\cos\phi} = \dfrac{-2.17 \text{ m / s}}{1.25 \text{ m / s}}$

Since $\tan\phi = \dfrac{\sin\phi}{\cos\phi}$, $\tan\phi = \dfrac{-2.17 \text{ m / s}}{1.25 \text{ m / s}}$

which yields two answers: $\phi = \tan^{-1}\left(\dfrac{-2.17}{1.25}\right) = -60°, \ 120°$

Looking at the original diagram, we can see that conservation of momentum would not allow the answer 120°. Therefore, we know that only the first is correct; the velocity of the second ball is 2.50 m/s, 60° below the $+x$-axis. $\Diamond$

(b) The total kinetic energy before collision is $KE_i = KE_{1i} + KE_{2i}$

or $KE_i = \dfrac{1}{2}m(5.00 \text{ m / s})^2 + 0 = m\left(12.5 \text{ m}^2/\text{s}^2\right)$

The total kinetic energy after collision is

$$KE_f = KE_{1f} + KE_{2f} = \frac{1}{2}m(4.33 \text{ m / s})^2 + m(2.50 \text{ m / s})^2$$

which yields $KE_f = m\left(12.5 \text{ m}^2/\text{s}^2\right)$

Since the total kinetic energy after collision is the same as that before collision, **the collision is elastic.** $\Diamond$

50. As shown in Figure P6.50, a bullet of mass m and speed v passes completely through a pendulum bob of mass M. The bullet emerges with a speed of $v/2$. The pendulum bob is suspended by a stiff rod of length ℓ and negligible mass. What is the minimum value of v such that the pendulum bob will barely swing through a complete vertical circle?

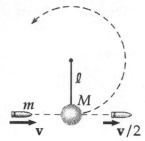

Figure P6.50

Solution   If the pendulum bob has **any** speed greater than zero when it reaches the top of the circular arc, it will pass through this highest point on the arc and swing down the other side of the arc to complete the full vertical circular motion.

Thus, the **minimum** speed it must have at the bottom of the circular arc is that which would allow it to reach the top with almost zero speed. To find this speed, apply conservation of energy over the interval that the bob is moving (after the bullet emerges from the bob). Conservation of **energy** is valid here since only a conservative force (gravity) performs work on the bob. Thus, choosing $y = 0$ at the bottom of the circular arc and V_{min} to be the bob's speed as the bullet exits,

$$KE_{bottom} + PE_{bottom} = KE_{top} + PE_{top} \quad \text{or} \quad \tfrac{1}{2}MV_{min}^{2} + 0 = 0 + Mg(2\ell)$$

This gives $V_{min} = \sqrt{4g\ell}$ as the minimum speed the bob must have just after the collision if it is to swing through a complete circle. To find the speed of the bullet before collision, apply the principle of conservation of **momentum**:

This gives $\quad p_f = p_i \qquad\qquad\qquad\qquad$ or $\quad m\left(\dfrac{v}{2}\right) + MV_{min} = mv + M(0)$

Therefore, $\quad MV_{min} = m(v/2) \qquad\qquad$ or $\quad v = 2MV_{min}/m$

Using the value found for V_{min} above, this yields the required speed of the bullet:

$$v = \frac{2M}{m}\sqrt{4g\ell} \qquad\qquad \text{or} \qquad v = 4\left(\frac{M}{m}\right)\sqrt{g\ell} \qquad\qquad \Diamond$$

54. A 12.0-g bullet is fired horizontally into a 100-g wooden block initially at rest on a horizontal surface. After impact, the block slides 7.5 m before coming to rest. If the coefficient of kinetic friction between block and surface is 0.650, what was the speed of the bullet immediately before impact?

Solution

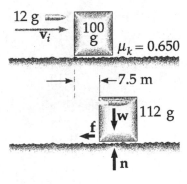

Unless a collision is perfectly elastic, it is extremely difficult to apply work-energy methods to any time interval that includes the collision. The collision in this case is perfectly **inelastic** since the block and bullet move as a single unit after collision. The strategy will be to employ conservation of **momentum** from just before the collision to just after the collision. Work-energy methods may be applied after the collision is over.

After the collision, the normal force exerted on the (block plus embedded bullet) by the surface is the weight of the block-bullet combination, or

$$n = w = (M+m)g = (0.100 \text{ kg} + 0.0120 \text{ kg})(9.80 \text{ m/s}^2) = 1.10 \text{ N}$$

The friction force that impedes the motion of the block-bullet combination is then

$$f = \mu_k n = 0.650(1.10 \text{ N}) = 0.713 \text{ N}$$

This is the only force doing work on the block-bullet combination as it slides 7.5 m and comes to rest. The normal force and weight are both perpendicular to the displacement and do no work.

Using the work-kinetic energy theorem, $W_{net} = KE_f - KE_i$, for the motion after collision gives:

$$f s \cos 180° = 0 - \tfrac{1}{2}(M+m)V^2: \qquad -(0.713 \text{ N})(7.5 \text{ m}) = -\tfrac{1}{2}(0.112 \text{ kg})V^2$$

Thus the velocity of the block and bullet just after the collision is $V = 9.8 \text{ m/s}$. Applying conservation of momentum from just before the collision to just after the collision gives: $p_i = p_f$. Substituting the appropriate terms,

$$m v_i + 0 = (M+m)V$$

The initial velocity of the bullet is then

$$v_i = \frac{(M+m)V}{m} = \frac{(0.112 \text{ kg})(9.8 \text{ m/s})}{0.012 \text{ kg}} = 91 \text{ m/s} \qquad \lozenge$$

59. A small block of mass $m_1 = 0.500$ kg is released from rest at the top of a curved wedge of mass $m_2 = 3.00$ kg, which sits on a frictionless horizontal surface as in Figure P6.59a. When the block leaves the wedge, its velocity is measured to be 4.00 m/s to the right, as in Figure P6.59b. (a) What is the velocity of the wedge after the block reaches the horizontal surface? (b) What is the height h of the wedge?

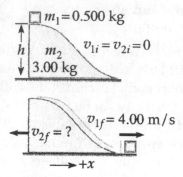

Figure P6.59 (a) and (b)

Solution

(a) Consider a system that consists of the wedge and block. The wedge sits on a frictionless horizontal surface, so the only **external** forces exerted on this system are gravitation forces (weight of wedge and weight of block) and the normal force exerted on the wedge by the horizontal surface. These **external forces** are all in the vertical direction. Since no horizontal external forces act on the system, the horizontal component of the system's momentum is constant:

$$\left(p_f\right)_x = \left(p_i\right)_x \qquad \text{or} \qquad m_1 v_{1f} + m_2 v_{2f} = m_1 v_{1i} + m_2 v_{2i}$$

This gives $v_{2f} = \dfrac{m_1\left(v_{1i} - v_{1f}\right) + m_2 v_{2i}}{m_2} = \dfrac{0.500 \text{ kg}(0 - 4.00 \text{ m / s}) - 0}{3.00 \text{ kg}}$

Thus, $v_{2f} = -0.667$ m / s $= 0.667$ m / s in the $-x$ direction   ◊

(b) To determine the height of the wedge, use work-energy techniques. Since no work is done on the block-wedge system by non-conservative forces, the total mechanical energy is constant. Choosing $y = 0$ at the horizontal surface,

$$\left(KE_{1f} + KE_{2f}\right) + \left(PE_{1f} + PE_{2f}\right) = \left(KE_{1i} + KE_{2i}\right) + \left(PE_{1i} + PE_{2i}\right)$$

or $\left(\tfrac{1}{2} m_1 v_{1f}^2 + \tfrac{1}{2} m_2 v_{2f}^2\right) + (0 + 0) = (0 + 0) + \left(m_1 g h + 0\right)$

$$\tfrac{1}{2}(0.500 \text{ kg})(4.00 \text{ m/s})^2 + \tfrac{1}{2}(3.00 \text{ kg})(0.667 \text{ m/s})^2 = (0.500 \text{ kg})(9.80 \text{ m/s}^2)h$$

Hence, the height of the wedge is $h = \dfrac{4.00 \text{ J} + 0.0667 \text{ J}}{4.90 \text{ N}} = 0.952$ m   ◊

63. A neutron in a reactor makes an elastic collision head on with a carbon atom that is initially at rest. (The mass of the carbon nucleus is about 12 times that of the neutron.) (a) What fraction of the neutron's kinetic energy is transferred to the carbon nucleus? (b) If the neutron's initial kinetic energy is 1.6×10^{-13} J, find its final kinetic energy and the kinetic energy of the carbon nucleus after the collision.

Solution Let particle 1 be the neutron and particle 2 be the carbon nucleus. Then, we are given that $m_2 = 12 m_1$.

(a) From conservation of momentum

$$m_2 v_{2f} + m_1 v_{1f} = m_1 v_{1i} + 0$$

Since $m_2 = 12 m_1$, this reduces to

$$12 v_{2f} + v_{1f} = v_{1i} \qquad [1]$$

For a head-on elastic collision,

$$v_{1i} - v_{2i} = -\left(v_{1f} - v_{2f}\right)$$

Since $v_{2i} = 0$, this becomes

$$v_{2f} - v_{1f} = v_{1i} \qquad [2]$$

Adding Equations [1] and [2] yields

$$13 v_{2f} = 2 v_{1i}$$

and

$$v_{2f} = \frac{2}{13} v_{1i}$$

The initial kinetic energy of the neutron is

$$KE_{1i} = \frac{1}{2} m_1 v_{1i}^2$$

and the final kinetic energy of the carbon nucleus is

$$KE_{2f} = \frac{1}{2} m_2 v_{2f}^2 = \frac{1}{2}(12 m_1)\left(\frac{4}{169} v_{1i}^2\right) = \frac{48}{169}\left(\frac{1}{2} m_1 v_{1i}^2\right) = \frac{48}{169} KE_{1i}$$

The fraction of kinetic energy transferred is

$$\frac{KE_{2f}}{KE_{1i}} = \frac{48}{169} = 0.28 \qquad \diamond$$

(b) If $KE_{1i} = 1.6 \times 10^{-13}$ J

then $KE_{2f} = \dfrac{48}{169} KE_{1i} = \dfrac{48}{169}\left(1.6 \times 10^{-13}\ \text{J}\right) = 4.5 \times 10^{-14}$ J $\qquad \diamond$

The kinetic energy remaining with the neutron is

$$KE_{1f} = KE_{1i} - KE_{2f} = 1.6 \times 10^{-13}\ \text{J} - 4.5 \times 10^{-14}\ \text{J} = 1.1 \times 10^{-13}\ \text{J} \qquad \diamond$$

Chapter 7

ROTATIONAL MOTION
AND THE LAW OF GRAVITY

NOTES ON SELECTED CHAPTER SECTIONS

7.1 Angular Speed and Angular Acceleration

Pure rotational motion refers to the motion of a rigid body about a fixed axis.

One **radian** (rad) is the angle subtended by an arc length equal to the radius of the arc. That is, the angle measured in radians is given by the arc length divided by the corresponding radius. One complete rotation of 360 degree equals 2π radians.

In the case of **rotation about a fixed axis**, every particle on the rigid body has the same angular velocity and the same angular acceleration.

7.2 Rotational Motion Under Constant Angular Acceleration

The equations for rotational motion under constant angular acceleration are of the **same form** as those for linear motion under constant linear acceleration with the substitutions $x \rightarrow \theta$, $v \rightarrow \omega$, and $a \rightarrow \alpha$.

7.3 Relations Between Angular and Linear Quantities

When a rigid body rotates about a fixed axis, every point in the object moves along a circular path which has its center at the axis of rotation. The **instantaneous velocity of each point is directed along a tangent to the circle**. Every point on the object experiences the same **angular speed**; however, points that are different distances from the axis of rotation have different tangential speeds. The value of each of the linear quantities [displacement (s), velocity (v), and acceleration (a_t)] is equal to the radial distance from the axis multiplied by the corresponding angular quantity, θ, ω, and α.

7.4 Centripetal Acceleration

In circular motion, the centripetal acceleration is directed inward toward the center of the circle and has a magnitude given either by v^2/r or $r\omega^2$.

7.6   Forces Causing Centripetal Acceleration

All **centripetal forces act toward the center of the circular path** along which the object moves. If the force that produces a centripetal acceleration vanishes, the object does not continue to move in its circular path; instead, it moves along a straight-line path tangent to the circle.

7.8   Newton's Law of Universal Gravitation

There are several important features of the law of universal gravitation:

1. The gravitational force is an **action-at-a distance force** that always exists between two particles regardless of the medium that separates them.

2. The force **varies as the inverse square of the distance between the particles** and therefore decreases rapidly with increasing separation.

3. The force is **proportional to the product of their masses.**

7.10   Kepler's Laws

Kepler's laws applied to the solar system are:

1. All planets move in elliptical orbits with the Sun at one of the focal points.

2. A line drawn from the Sun to any planet sweeps out equal areas in equal time intervals.

3. The square of the orbital period of any planet is proportional to the cube of the average distance from the planet to the Sun.

EQUATIONS AND CONCEPTS

When a particle moves along a circular path of radius r, the distance traveled by the particle is called the arc length, s. The radial line from the center of the path to the particle sweeps out an angle, θ.

$$\theta \equiv \frac{s}{r} \qquad (7.1)$$

where $\theta\,(\text{rad}) = \left(\dfrac{\pi}{180°}\right)\!\left(\theta\,(\text{deg})\right)$

The angle θ in Equation 7.1 is the ratio of two lengths (arc length to radius) and hence is a **dimensionless quantity**. It is common practice to refer to the angle as being in units of radians. **The angle in Equation 7.1 must have units of radians.**

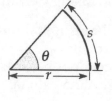

$$1 \text{ rad} = \left(\frac{360°}{2\pi}\right) = 57.3°$$

The average angular velocity of a rotating object is the ratio of the angular displacement to the time interval during which the angular displacement occurs.

$$\overline{\omega} = \frac{\theta_f - \theta_i}{t_f - t_i} = \frac{\Delta\theta}{\Delta t} \qquad (7.2)$$

The average angular acceleration of a rotating object is the ratio of change in angular velocity to the time interval during which the change in velocity occurs.

$$\overline{\alpha} = \frac{\omega_f - \omega_i}{t_f - t_i} = \frac{\Delta\omega}{\Delta t} \qquad (7.4)$$

To the right are given the equations of rotational motion with constant acceleration and the corresponding equations for linear motion with constant acceleration. Note that the rotational equations, involving the angular variables $\Delta\theta$, ω, and α, have a one-to-one correspondence with the equations of linear motion, involving the variables Δx, v, and a.

$$\omega = \omega_i + \alpha t \qquad (7.5)$$

$$v = v_i + at$$

$$\Delta\theta = \omega_i t + \frac{1}{2}\alpha t^2 \qquad (7.6)$$

$$\Delta x = v_i t + \frac{1}{2}at^2$$

$$\omega^2 = \omega_i{}^2 + 2\alpha\,\Delta\theta \qquad (7.7)$$

$$v^2 = v_i{}^2 + 2a\Delta x$$

The tangential velocity and tangential acceleration of a given point on a rotating object are related to the corresponding angular quantities via the radius of the circular path along which the point moves.

$$v_t = r\omega \qquad (7.8)$$

$$a_t = r\alpha \qquad (7.9)$$

The tangential velocity and tangential acceleration are along directions which are tangent to the circular path (and therefore perpendicular to the radius from the center of rotation). Also note that **every point on a rotating object has the same value of ω and the same value of α.** However, points which are at different distances from the axis of rotation have different values of v_t and a_t.

Comment on rotational velocity and acceleration

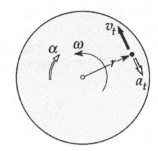

In the diagram above, a rigid disk is rotating counterclockwise $(+\omega)$ about an axis through its center with a decreasing rate of rotation $(-\alpha)$.

An object in circular motion has a centripetal acceleration which is **directed toward the center of the circle** and has a magnitude which depends on the values of the tangential velocity and the radius of the path.

$$a_c = \frac{v^2}{r} \qquad (7.11)$$

or

$$a_c = r\omega^2 \qquad (7.14)$$

An object which is moving along a circular path with increasing or decreasing speed has both a tangential component of acceleration (tangent to the instantaneous direction of travel) and a centripetal component of acceleration (perpendicular to the instantaneous direction of travel). The magnitude and direction of the total acceleration can be found by the usual methods of **vector addition**.

$$a = \sqrt{a_t^2 + a_c^2} \qquad (7.15)$$

$$\theta = \tan^{-1}\left(\frac{a_t}{a_c}\right)$$

Forces which maintain motion along a circular path are directed toward the center of the path and are called **centripetal forces**. These forces are also called radial forces.

$$F_c = ma_c = m\frac{v_t^2}{r} \qquad (7.16)$$

The law of universal gravitation states that every particle in the Universe attracts every other particle with a force that is directly proportional to the product of their masses and inversely proportional to the square of the distance between them. The constant G is called the universal gravitational constant.

$$F = G\frac{m_1 m_2}{r^2} \tag{7.17}$$

$$G = 6.673 \times 10^{-11} \ \text{N} \cdot \text{m}^2/\text{kg}^2 \tag{7.18}$$

The gravitational force between two masses m_1 and m_2 is one of attraction; each mass exerts a force of attraction on the other. These two forces form an action-reaction pair. **The magnitudes of the forces of gravitational attraction on the two masses are equal regardless of the relative values of m_1 and m_2.** This is in agreement with Newton's third law (action - reaction).

Comments on the gravitational force.

$\mathbf{F}_{12} \quad \mathbf{F}_{21}$
$m_1 \qquad m_2$

The escape speed of an object projected upward from the Earth's surface is independent of the mass of the object.

$$v_{\text{esc}} = \sqrt{\frac{2GM_E}{R_E}} \tag{7.21}$$

Kepler's third law states that the square of the orbital period of a planet is proportional to the cube of the mean distance from the planet to the Sun. For an Earth satellite, M_S in Equation 7.22 must be replaced by M_E, the mass of the Earth.

$$T^2 = \left(\frac{4\pi^2}{GM_S}\right)r^3 = K_S r^3 \tag{7.22}$$

K_S is independent of the mass of the planet.

$$K_S = \frac{4\pi^2}{GM_S} = 2.97 \times 10^{-19} \ \text{s}^2/\text{m}^3$$

SUGGESTIONS, SKILLS, AND STRATEGIES

The following features involving centripetal forces and centripetal accelerations should be kept in mind:

1. Draw a free-body diagram of the object(s) under consideration, showing all forces that act on it (them).

2. Choose a coordinate system with one axis tangent to the path followed by the object and the other axis perpendicular to the plane of the circular path.

3. Find the net force toward the center of the circular path. This is the centripetal force, the force which causes the centripetal acceleration.

4. From this point onward, the steps are virtually identical to those encountered when solving Newton's second law problems with $F_x = ma_x$ and $F_y = ma_y$. In this case, Newton's second law is applied along the radial (directed toward the center) and tangential directions. Also, you should note that the magnitude of the centripetal acceleration can always be written as $a_c = v^2/r$.

REVIEW CHECKLIST

You should understand that:

▷ Quantitatively, the angular displacement, angular velocity, and angular acceleration for a rigid body system in rotational motion are related to the distance traveled, tangential velocity, and tangential acceleration, respectively. The linear quantity is calculated by multiplying the angular quantity by the radius arm for an object or point in that system.

▷ If a body rotates about a fixed axis, every particle on the body has the same angular velocity and angular acceleration. For this reason, rotational motion can be simply described using these quantities. The formulas which describe angular motion are analogous to the corresponding set of equations pertaining to linear motion.

▷ The nature of the acceleration of a particle moving in a circle with **constant speed** is such that, although **v** = constant, the **direction** of **v** varies in time, which is the origin of the radial, or centripetal acceleration.

▷ When both the magnitude and direction of **v** are changing with time, there are two components of acceleration for a particle moving on a curved path. In this case, the particle has a tangential component of acceleration and a radial component of acceleration.

▷ Newton's law of universal gravitation is an example of an inverse-square law, and it describes an **attractive** force between two **particles** separated by a distance **r**.

SOLUTIONS TO SELECTED END-OF-CHAPTER PROBLEMS

3. Find the angular speed of Earth about the Sun in radians per second and degrees per day.

Solution The Earth completes a full orbit (360° or 2π radians) around the Sun in one year $(3.156 \times 10^7 \text{ s})$. Therefore, the average angular speed of the Earth's orbital motion is

$$\bar{\omega} = \frac{\Delta\theta}{\Delta t} = \frac{2\pi \text{ rad}}{3.156 \times 10^7 \text{ s}} = 1.991 \times 10^{-7} \text{ rad / s} \qquad \diamond$$

Converting to units of degrees per day yields

$$\bar{\omega} = \left(1.991 \times 10^{-7} \text{ rad / s}\right)\left(\frac{57.3 \text{ deg}}{1 \text{ rad}}\right)\left(\frac{8.64 \times 10^4 \text{ s}}{1 \text{ day}}\right) = 0.986 \text{ deg / day} \qquad \diamond$$

7. A car is traveling with a velocity of 17.0 m/s down a straight horizontal highway. The wheels of the car have a radius of 48.0 cm. If the car then speeds up with an acceleration of 2.00 m / s^2 for 5.00 s, find the number of revolutions of the wheels during this period.

Solution The linear velocity of the car at the end of the 5.00-s interval is

$$v_f = v_i + at = 17.0 \text{ m / s} + \left(2.00 \text{ m / s}^2\right)(5.00 \text{ s}) = 27.0 \text{ m / s}$$

The average velocity of the car is

$$\bar{v} = \frac{v_i + v_f}{2} = \frac{17.0 \text{ m / s} + 27.0 \text{ m / s}}{2} = 22.0 \text{ m / s}$$

Its linear displacement during this time is $\qquad s = \bar{v}t = (22.0 \text{ m / s})(5.00 \text{ s}) = 110 \text{ m}$

If the car's wheels roll without slipping on the highway, the relation between the linear displacement (s) of the center of a wheel and the angular displacement (θ) of the wheel is $s = r\theta$ where r is the radius of the wheel. Thus, $\theta = s/r = 110 \text{ m}/0.48 \text{ m} = 229 \text{ rad}$.

This may be converted to revolutions as: $\qquad \theta = (229 \text{ rad})\left(\frac{1 \text{ rev}}{2\pi \text{ rad}}\right) = 36.5 \text{ rev} \qquad \diamond$

9. The diameters of the main rotor and tail rotor of a single-engine helicopter are 7.60 m and 1.02 m, respectively. The respective rotational speeds are 450 rev/min and 4138 rev/min. Calculate the speeds of the tips of both rotors. Compare these speeds with the speed of sound, 343 m/s.

Solution The tips of one of the rotors move in a circular path with the specified angular velocity (or rotational speed), ω, for that rotor. The linear speed of these rotor tips is given by $v = r\omega$, where r is the radius of the circular path.

For the main rotor: $\quad v = r\omega = \left(\dfrac{7.60 \text{ m}}{2}\right)\left(450 \ \dfrac{\text{rev}}{\text{min}}\right)\left(\dfrac{1 \text{ min}}{60 \text{ s}}\right)\left(\dfrac{2\pi \text{ rad}}{1 \text{ rev}}\right) = 179 \text{ m/s} \qquad \Diamond$

and $\quad v = (179 \text{ m/s})\left(\dfrac{v_{\text{sound}}}{343 \text{ m/s}}\right) = 0.522 \, v_{\text{sound}} \qquad \Diamond$

For the tail rotor: $\quad v = r\omega = \left(\dfrac{1.02 \text{ m}}{2}\right)\left(4138 \ \dfrac{\text{rev}}{\text{min}}\right)\left(\dfrac{1 \text{ min}}{60 \text{ s}}\right)\left(\dfrac{2\pi \text{ rad}}{1 \text{ rev}}\right) = 221 \text{ m/s} \qquad \Diamond$

and $\quad v = (221 \text{ m/s})\left(\dfrac{v_{\text{sound}}}{343 \text{ m/s}}\right) = 0.644 \, v_{\text{sound}} \qquad \Diamond$

---

14. It has been suggested that rotating cylinders about 10 mi long and 5.0 mi in diameter be placed in space and used as colonies. What angular speed must such a cylinder have so that the centripetal acceleration at its surface equals the free-fall acceleration?

Solution If an object moves along a circular path of radius r with a tangential speed v_t, it is always accelerating toward the center of the circular path with an acceleration of

$$a_c = v_t^2 / r = r\omega^2$$

It is desired that $\quad a_c = g_{\text{earth}} = 9.80 \text{ m/s}^2 \quad$ for an object on the surface of a rotating cylinder that has a radius

$$r = \tfrac{1}{2}(\text{diameter}) = 2.5 \text{ mi}$$

The required angular speed, ω, is then:

$$\omega = \sqrt{\dfrac{a_c}{r}} = \sqrt{\dfrac{9.80 \text{ m/s}^2}{2.5 \text{ mi}(1609 \text{ m/1 mi})}} = 4.9 \times 10^{-2} \text{ rad/s} \qquad \Diamond$$

---

23. A 50.0-kg child stands at the rim of a merry-go-round of radius 2.00 m, rotating with an angular speed of 3.00 rad / s. (a) What is the child's centripetal acceleration? (b) What is the minimum force between his feet and the floor of the carousel that is required to keep him in the circular path? (c) What minimum coefficient of static friction is required? Is the answer you found reasonable? In other words, is he likely to be able to stay on the merry-go-round?

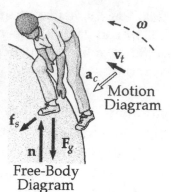

Motion Diagram

Free-Body Diagram

Solution

(a) When the child stands, without slipping, at the rim of the rotating merry-go-round, he follows a horizontal circular path of radius $r = 2.00$ m with an angular speed, $\omega = 3.00$ rad / s. His centripetal acceleration is

$$a_c = \frac{v_t^2}{r} = \frac{(r\omega)^2}{r} = r\omega^2 = (2.00 \text{ m})(3.00 \text{ rad/s})^2 = 18.0 \text{ m/s}^2 \qquad \Diamond$$

(b) The force, directed toward the center of the circular path, required to produce this centripetal acceleration is $F_c = ma_c$ or

$$F_c = (50.0 \text{ kg})(18.0 \text{ m/s}^2) = 900 \text{ N} \qquad \Diamond$$

(c) Three forces act on the child as he stands on the rotating merry-go-round. These are: (1) a normal force, n, exerted upward on the child by the merry-go-round; (2) a downward gravitational force

$$F_g = w = mg = (50.0 \text{ kg})(9.80 \text{ m/s}^2) = 490 \text{ N}$$

and (3) a horizontal friction force between his feet and the platform. Since the child's vertical acceleration is zero, Newton's second law gives

$$\sum F_y = n - F_g = 0 \qquad \text{or} \qquad n = F_g = 490 \text{ N}$$

The only force that is horizontal and thus capable of producing an acceleration toward the center of the horizontal circular path is the static friction force. Therefore, it is necessary that $f_s = F_c$. Since $f_s \leq \mu_s n$, this means that $F_c \leq \mu_s n$, and the required coefficient of static friction is

$$\mu_s \geq \frac{F_c}{n} = \frac{900 \text{ N}}{490 \text{ N}} = 1.84$$

Comparing this result to common values for μ_s (see Table 4.2 in the textbook), this required value for the coefficient of static friction is seen to be unrealistic. The child will not be able to stay on the merry-go-round. $\qquad \Diamond$

26. Tarzan ($m = 85$ kg) tries to cross a river by swinging from a 10-m-long vine. His speed at the bottom of the swing (as he just clears the water) is 8.0 m/s. Tarzan doesn't know that the vine has a breaking strength of 1000 N. Does he make it safely across the river? Justify your answer.

Solution   When Tarzan is at the bottom of the swing, the vine is vertical with its tension force, T, acting upward on Tarzan. The only other force present is his weight, $F_g = mg$, directed downward. Therefore, the net force directed toward the center of Tarzan's vertical circular path is $F_{net} = T - F_g$.

If Tarzan is to stay on the circular path, it is necessary that

$F_{net} = ma_c = mv_t^2/r$.

This gives $T - F_g = m\dfrac{v_t^2}{r}$

or $T = mg + m\dfrac{v_t^2}{r} = \left(85 \text{ kg}\right)\left(9.80 \text{ m / s}^2 + \dfrac{\left(8.0 \text{ m / s}\right)^2}{10 \text{ m}}\right) = 1.4 \times 10^3 \text{ N}$

Thus, the required tension in the vine exceeds the breaking strength of 1000 N. We must conclude that Tarzan will not cross safely.   ◊

29.   The average distance separating Earth and the Moon is 384 000 km. Use the data in Table 7.3 to find the net gravitational force exerted by Earth and the Moon on a spaceship with mass 3.00×10^4 kg located halfway between them.

Solution   When a spaceship (mass m_s) is halfway between the Earth (mass m_E) and the Moon (mass m_M), it is at a distance

$$r = \left(192\,000 \text{ km}\right)\left(\dfrac{1.00 \times 10^3 \text{ m}}{1 \text{ km}}\right) = 1.92 \times 10^8 \text{ m}$$

from each body. From Newton's Universal Law of Gravitation, the forces exerted on the ship by Earth and by the Moon are

$F_E = Gm_E m_s/r^2$ (directed toward Earth)

and $F_M = Gm_M m_s/r^2$ (directed toward the Moon.)

105

From Table 7.3 in the textbook, $m_E = 5.98 \times 10^{24}$ kg and $m_M = 7.36 \times 10^{22}$ kg. Since $F_E > F_M$, the net gravitational force acting on the ship is directed toward the Earth and has a magnitude of:

$$F_{net} = F_E - F_M = \frac{Gm_E m_s}{r^2} - \frac{Gm_M m_s}{r^2} = \frac{Gm_s(m_E - m_M)}{r^2}$$

or $\quad F_{net} = \dfrac{\left(6.67 \times 10^{-11} \text{ N} \cdot \text{m}^2 / \text{kg}^2\right)\left(3.00 \times 10^4 \text{ kg}\right)\left(5.98 \times 10^{24} \text{ kg} - 7.36 \times 10^{22} \text{ kg}\right)}{\left(1.92 \times 10^8 \text{ m}\right)^2}$

This yields $F_{net} = 321$ N directed toward the Earth. ◊

---

35. A satellite moves in a circular orbit around Earth at a speed of 5000 m/s. Determine (a) the satellite's altitude above Earth's surface and (b) the period of the satellite's orbit.

Solution

(a) The gravitational force exerted on the satellite (mass m) by the Earth (mass m_E) must produce the required centripetal acceleration, $a_c = v_t^2/r$, of the satellite. That is:

$$\frac{Gm_E m}{r^2} = m\left(\frac{v_t^2}{r}\right) \qquad \text{which reduces to} \qquad r = \frac{Gm_E}{v_t^2}$$

The radius of the satellite's orbit is therefore

$$r = \left(6.67 \times 10^{-11} \text{ N} \cdot \text{m}^2 / \text{kg}^2\right)\frac{\left(5.98 \times 10^{24} \text{ kg}\right)}{\left(5000 \text{ m} / \text{s}\right)^2} = 1.60 \times 10^7 \text{ m}$$

The altitude above the surface of the Earth, of radius R_E, is then

$$h = r - R_E = 1.60 \times 10^7 \text{ m} - 6.37 \times 10^6 \text{ m} = 9.58 \times 10^6 \text{ m} \qquad ◊$$

(b) The period is the time required for the satellite to complete one full trip around the circular orbit. This is:

$$T = \frac{\text{circumference of orbit}}{\text{orbital speed}} = \frac{2\pi r}{v_t}$$

$$T = \frac{2\pi\left(1.60 \times 10^7 \text{ m}\right)}{5000 \text{ m} / \text{s}} = 2.00 \times 10^4 \text{ s} = \left(2.00 \times 10^4 \text{ s}\right)\left(\frac{1 \text{ h}}{3600 \text{ s}}\right) = 5.57 \text{ h} \qquad ◊$$

37. Io, a satellite of Jupiter, has an orbital period of 1.77 days and an orbital radius of 4.22×10^5 km. From these data, determine the mass of Jupiter.

Solution   The force available to produce the needed centripetal acceleration, a_c, and hold Io in its orbit is the gravitational force exerted on it by Jupiter. Thus, Newton's second law gives the needed magnitude of this force as

$$F_g = m_I a_c = \frac{m_I v_t^2}{r}$$

where m_I is the mass of Io, v_t is its orbital speed, and r is its orbital radius. But, Newton's Universal Law of Gravitation gives the magnitude of this gravitational force as $F_g = Gm_J m_I / r^2$ where m_J is the mass of Jupiter. Therefore, it necessary that $Gm_J m_I / r^2 = m_I v_t^2 / r$. Note that the mass of the satellite (Io in this case) cancels and the mass of the central body (Jupiter in this case) is given as:

$$m_J = \frac{r v_t^2}{G} \qquad [1]$$

The radius of Io's orbit is $\qquad r = 4.22 \times 10^5 \text{ km} = 4.22 \times 10^8 \text{ m}$,

and its orbital period is $\qquad T = 1.77 \text{ days}(86\,400 \text{ s / day}) = 1.53 \times 10^5 \text{ s}$

The orbital speed of Io is then

$$v_t = \frac{\text{circumference}}{\text{period}} = \frac{2\pi r}{T} = \frac{2\pi\left(4.22 \times 10^8 \text{ m}\right)}{1.53 \times 10^5 \text{ s}} = 1.73 \times 10^4 \text{ m / s}$$

Then, the mass of Jupiter must be:

$$m_J = \frac{\left(4.22 \times 10^8 \text{ m}\right)\left(1.73 \times 10^4 \text{ m / s}\right)^2}{6.67 \times 10^{-11} \text{ N} \cdot \text{m}^2 / \text{kg}^2} = 1.90 \times 10^{27} \text{ kg} \qquad \diamond$$

Note: This problem illustrates a very useful tool with which astronomers can determine the mass of celestial bodies having observable satellites.

47. A car moves at speed v across a bridge made in the shape of a circular arc of radius r. (a) Find an expression for the normal force acting on the car when it is at the top of the arc. (b) At what minimum speed will the normal force become zero (causing occupants of the car to seem weightless) if $r = 30.0$ m?

Solution Consider the sketch at the right showing the forces acting on the car as it crosses the top of the arc of the bridge.

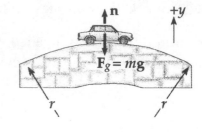

(a) If the car remains in contact with the bridge, it follows a circular path of radius r. Thus, it must be accelerating **toward the center** of the circular path at a rate $a_c = v^2/r$.

When the car is at the top of the arc, the line toward the center is vertical. Taking upward as positive and applying Newton's second law to the vertical motion,

$$\sum F_y = ma_y \qquad \text{yields} \qquad n - mg = m(-a_c) \qquad \text{or} \qquad n = m\left(g - \frac{v^2}{r}\right) \qquad \Diamond$$

(b) If the occupants of the car (and the car itself) are to seem weightless, the downward force they exert on the roadway must be zero. Then, by Newton's third law, the upward force the roadway exerts on them (i.e., the normal force n) must also be zero. Therefore, the desired speed is that for which

$$n = m\left(g - v^2/r\right) = 0 \qquad \text{or} \qquad v = \sqrt{rg}$$

If the radius of the circular arc is 30.0 m, then

$$v = \sqrt{(30.0 \text{ m})(9.80 \text{ m}/\text{s}^2)} = 17.1 \text{ m}/\text{s} \qquad \Diamond$$

---

51. In a popular amusement park ride, a rotating cylinder of radius of radius 3.00 m is set in rotation at an angular speed of 5.00 rad/s, as in Figure P7.51. The floor then drops away, leaving the riders suspended against the wall in a vertical position. What minimum coefficient of friction between a rider's clothing and the wall is needed to keep the rider from slipping? (**Hint:** Recall that the magnitude of the maximum force of static friction is equal to μn, where n is the normal force — in this case, the force causing the centripetal acceleration.)

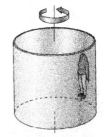

Figure P7.51

Solution The normal force exerted on the person by the cylindrical wall must provide the centripetal acceleration, so

$$n = m\left(\frac{v_t^2}{r}\right) = \frac{m(r\omega)^2}{r} = m\left(r\omega^2\right)$$

If the minimum acceptable coefficient of friction is present, the person is on the verge of slipping and the maximum static friction force equals the person's weight, or

$$(f_s)_{max} = (\mu_s)_{min} n = mg$$

Thus, $\quad (\mu_s)_{min} = \dfrac{mg}{n} = \dfrac{g}{r\omega^2} = \dfrac{9.80 \text{ m / s}^2}{(3.00 \text{ m})(5.00 \text{ rad / s})^2} = 0.131$ ◊

56. Show that the escape speed from the surface of a planet of uniform density is directly proportional to the radius of the planet.

Solution Some material required for the solution of this problem is found in Section 7.9 (**Gravitational Potential Energy Revisited**) of the textbook. If an object of mass m is fired vertically at speed v_e from the surface of a planet of mass M_p and radius R_p, its total mechanical energy is

$$E_i = KE_i + PE_i = \frac{1}{2}mv_e^2 - \frac{GM_p m}{R_p} : \qquad E_f = KE_f + PE_f = KE_i + PE_i = E_i$$

If the object is to escape from the planet, it must reach a location where the gravitational force exerted on it by the planet $F = GM_p m/r^2$ is zero. This is true only when $r \to \infty$. But when $r \to \infty$, the gravitational potential energy of the object is

$$PE_f = -\frac{GM_p m}{r} \to -\frac{GM_p m}{\infty} = 0$$

Also, if the object leaves the planet's surface with the minimum required speed to reach this location ($r \to \infty$), it will arrive with zero final speed, $v_f = 0$.

Thus, $\quad KE_f = \frac{1}{2}mv_f^2 = 0$, giving $\quad E_f = KE_f + PE_f = 0$

Since $\quad E_i = E_f = \text{constant}$, $\qquad E_i = \frac{1}{2}mv_e^2 - GM_p m / R_p = 0$,

and the escape velocity is $\qquad v_e = \sqrt{2GM_p/R_p}$

If the planet has uniform density, its mass is $\quad M_p = \text{density} \times \text{volume} = \rho\left(\frac{4}{3}\pi R_p^3\right)$ and the escape velocity becomes

$$v_e = \sqrt{(2G/R_p)\rho\left(\tfrac{4}{3}\pi R_p^3\right)} = \sqrt{\tfrac{8}{3}\pi\rho GR_p^2} = R_p\sqrt{\tfrac{8}{3}\pi\rho G} = (R_p)(\text{constant})$$

Thus, the escape velocity is directly proportional to the planet radius, R_p ◊

Chapter 8
ROTATIONAL EQUILIBRIUM AND ROTATIONAL DYNAMICS

NOTES ON SELECTED CHAPTER SECTIONS

8.1 Torque

Torque is the physical quantity which is a measure of the tendency of a force to cause rotation of a body about a specified axis. It is important to remember that **torque must be defined with respect to a specific axis of rotation**. Torque, which has the SI **units** of N·m, must not be confused with work energy.

8.2 Torque and the Two Conditions For Equilibrium

A body in static equilibrium must satisfy two conditions:

1. The resultant external force must be zero.
2. The resultant external torque must be zero.

You should note that it does not matter where you pick the axis of rotation for calculating the net torque if the object is in equilibrium; since the object is not rotating, the location of the axis is completely arbitrary.

8.3 The Center of Gravity

In order to calculate the torque due to the weight (gravitational force) on a rigid body, the entire weight of the object can be considered to be concentrated at a single point called the center of gravity. The center of gravity of a homogeneous, symmetric body must lie along an axis of symmetry.

8.5 Relationship Between Torque and Angular Acceleration

The angular acceleration of an object is proportional to the net torque acting on it. The moment of inertia of the object is the proportionality constant between the net torque and the angular acceleration. **The force and mass in linear motion correspond to torque and moment of inertia in rotational motion.** Moment of inertia of an object depends on the location of the axis of rotation and upon the manner in which the mass is distributed relative to that axis (e.g., a ring has a greater moment of inertia than a disk of the same mass and radius).

8.6 Rotational Kinetic Energy

In linear motion, the energy concept is useful in describing the motion of a system. The energy concept can be equally useful in simplifying the analysis of

110

rotational motion. We now have expressions for four types of mechanical energy: gravitational potential energy, PE_g; elastic potential energy, PE_s, translational kinetic energy, KE_t; and rotational kinetic energy, KE_r. We must include all these forms of energy in the equation for potential conservation of mechanical energy.

EQUATIONS AND CONCEPTS

This is the first condition for equilibrium. An object will remain at rest or move with uniform motion along a straight line when no net external force acts on it. This condition corresponds to **translational equilibrium**. The vector equation for translational equilibrium can be written in component form (in this case for a two-dimensional situation).

$$\sum \mathbf{F} = 0$$

$$\sum F_x = 0$$

$$\sum F_y = 0$$

A set of two or more forces are concurrent if their extended lines of action intersect at a single point (as shown in the diagram at right). When **concurrent** forces act upon an object, the condition $\Sigma \mathbf{F} = 0$ is sufficient to ensure equilibrium. **Review the recommended solution procedure and example problems in the textbook.**

Comment on
concurrent forces.

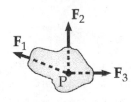

When nonconcurrent forces act on a body, the first condition ($\Sigma \mathbf{F} = 0$) is not sufficient to ensure complete equilibrium. In this case it is necessary to consider the net torque acting on the body relative to some axis. For a given force, the magnitude of the corresponding **torque is the product of the magnitude of the force and the lever arm**. Remember that the moment arm is the perpendicular distance from the axis of rotation to the line along which the force is acting.

$$\tau = Fd \qquad (8.1)$$

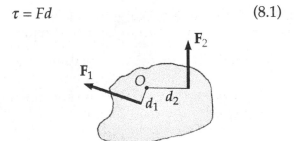

Note: For this object,
$$\tau_1 = -F_1 d_1 \quad \text{and} \quad \tau_2 = +F_2 d_2$$

111

The sign of a torque due to a force is considered positive if the force has a tendency to rotate the body counterclockwise about the chosen axis, and negative if the tendency for rotation is clockwise.

Comment on sign conventions

The first condition for mechanical equilibrium requires that the net external force acting on a body equal zero. This will ensure that the body is in translational equilibrium. The second condition for mechanical equilibrium requires that the net torque acting on a body equals zero. This condition will ensure that the body is in **rotational equilibrium**.

$$\sum \mathbf{F} = 0$$

$$\sum \tau = 0 \tag{8.2}$$

To compute the torque due to the force of gravity (an object's weight), the total weight can be considered as being concentrated at a single point called the center of gravity and having coordinates x_{cg}, y_{cg}.

$$x_{cg} = \frac{\sum m_i x_i}{\sum m_i} \tag{8.3}$$

$$y_{cg} = \frac{\sum m_i y_i}{\sum m_i} \tag{8.4}$$

The angular acceleration of a point mass, moving in a path of radius r, is proportional to the net torque acting on the mass.

$$\tau = mr^2 \alpha \tag{8.6}$$

For a **point mass** m, moving in a path of radius r, $I = mr^2$.

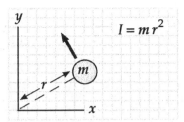

For a collection of discrete point masses, each with its own corresponding value of m and r, $I = \sum_i m_i r_i^2$.

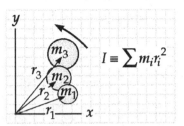

The moment of inertia of a rigid body depends on the **mass and the distribution of mass relative to the axis of rotation.**

$$I \equiv \sum mr^2 \qquad (8.8)$$

The value of I can be calculated for an extended object with good symmetry. Expressions for the moments of inertia for a number of objects of common shape are given in Table 8.1 of your text. The SI units of moment of inertia are kg·m².

The angular acceleration of an extended object is proportional to the net torque acting on the object. The proportionality constant, I, is called the moment of inertia of the object.

$$\sum \tau = I\alpha \qquad (8.9)$$

A rigid body or mass in rotational motion has kinetic energy due to its motion. Note that this is not a new form of energy but is a convenient form for representing kinetic energy associated with rotational motion.

$$KE_r = \frac{1}{2} I \omega^2 \qquad (8.11)$$

When only conservative forces act on a system, the total mechanical energy (gravitational potential, rotational kinetic, and translational kinetic) of the system is conserved.

$$\left(KE_t + KE_r + PE_g + PE_s\right)_i = \left(KE_t + KE_r + PE_g + PE_s\right)_f \qquad (8.12)$$

An object in rotational motion is characterized by a quantity, L, which is called angular momentum. **Angular momentum is a vector quantity** and has the direction of the angular velocity.

$$L \equiv I \omega \qquad (8.13)$$

The net external torque acting on an object equals the time rate of change of its angular momentum. Note that this is the **rotational analog of Newton's second law for transitional motion**.

$$\tau = \frac{\Delta L}{\Delta t} \qquad (8.14)$$

When the net external torque acting on a system is zero, the angular momentum of the system is conserved.

$$L_i = L_f \qquad \text{if} \qquad \Sigma \tau = 0, \qquad (8.15)$$

$$I_i \omega_i = I_f \omega_f \qquad \text{if} \qquad \Sigma \tau = 0 \qquad (8.16)$$

SUGGESTIONS, SKILLS, AND STRATEGIES

PROBLEM-SOLVING STRATEGY FOR OBJECTS IN EQUILIBRIUM

1. Draw a simple, neat diagram of the system that is large enough to show all the forces clearly.

2. Isolate the object of interest being analyzed. Draw a free-body diagram for this object showing all external forces acting on the object. For systems containing more than one object, draw separate diagrams for each object. Do not include forces that the object exerts on its surroundings.

3. Establish convenient coordinate axes for each body and find the components of the forces along these axes. Now apply the first condition of equilibrium for each object under consideration; namely, that the net force on the object in the x and y directions must be zero.

4. Choose a convenient origin for calculating the net torque on the object. Now apply the second condition of equilibrium that says that the net torque on the object about any origin must be zero. Remember that the choice of the origin for the torque equation is arbitrary; therefore, choose an origin that will simplify your calculation as much as possible. Note that a force that acts along a line passing through the point chosen as the axis of rotation gives zero contribution to the torque.

5. The first and second conditions for equilibrium will give a set of simultaneous equations with several unknowns. To complete your solution, all that is left is to solve for the unknowns in terms of the known quantities.

PROBLEM-SOLVING STRATEGY FOR ROTATIONAL MOTION

The following facts and procedures should be kept in mind when solving rotational motion problems.

1. Problems involving the equation $\Sigma\tau = I\alpha$ are very similar to those encountered in Newton's second law problems, $\Sigma\mathbf{F} = m\mathbf{a}$. Note the correspondences between linear and rotational quantities in that $\mathbf{F}$ is replaced by τ, m by I, and a by α.

2. Other analogues between rotational quantities and linear quantities include the replacement of x by θ and v by ω. Recall that each linear quantity (x, v, and a)

equals the product of the radius and the corresponding angular quantity (θ, ω, and α). These are helpful as memory devices for such rotational motion quantities as rotational kinetic energy, $KE_r = \frac{1}{2}I\omega^2$, and angular momentum, $L = I\omega$.

3. With the analogues mentioned in Step 2, conservation of energy techniques remain the same as those examined in Chapter 5, except for the fact that a new kind of energy, rotational kinetic energy, must be included in the expression for the conservation of energy (see Equation 8.12).

4. Likewise, the techniques for solving conservation of angular momentum problems are essentially the same as those used in solving conservation of linear momentum problems, except you are equating total angular momentum before to total angular momentum after as $I_i\omega_i = I_f\omega_f$.

REVIEW CHECKLIST

▷ There are two necessary conditions for equilibrium of a rigid body: $\Sigma F = 0$ and $\Sigma \tau = 0$. Torques which cause counterclockwise rotations are positive and those causing clockwise rotations are negative.

▷ The torque associated with a force has a magnitude equal to the force times the lever arm. The lever arm is the perpendicular distance from the axis of rotation to a line drawn along the direction of the force. Also, the net torque on a rigid body about some axis is proportional to the angular acceleration; that is, $\tau = I\alpha$, where I is the moment of inertia about the axis about which the net torque is evaluated.

▷ The work-energy theorem can be applied to a rotating rigid body. That is, the net work done on a rigid body **rotating about a fixed axis** equals the change in its rotational kinetic energy. The **law of conservation of mechanical energy** can be used in the solution of problems involving rotating rigid bodies.

▷ The time rate of change of the angular momentum of a rigid body rotating about an axis is proportional to the net torque acting about the axis of rotation. This is the rotational analog of Newton's second law.

SOLUTIONS TO SELECTED END-OF-CHAPTER PROBLEMS

5. A simple pendulum consists of a small object of mass 3.0 kg hanging at the end of a 2.0-m-long light string that is connected to a pivot point. Calculate the magnitude of the torque (due to the force of gravity) about this pivot point when the string makes a 5.0° angle with the vertical.

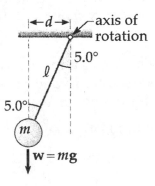

Solution The torque due to a force of magnitude F is $\tau = Fd$, where the lever arm d is the perpendicular distance from the axis of rotation to the line drawn along the direction of the force. Consider the force of gravity (i.e., the weight $w = mg$) acting on the bob of a simple pendulum as shown in the sketch at the right. From the sketch, the lever arm, d, of the force **w** is $d = \ell \sin 5.0°$, where ℓ is the length of the pendulum string. Thus, the torque of this force about the indicated rotation axis is $\tau = wd = mg\ell \sin 5.0°$. Using the given data, this becomes

$$\tau = (3.0 \text{ kg})(9.8 \text{ m / s}^2)(2.0 \text{ m})\sin 5.0° \qquad \text{or} \qquad \tau = 5.1 \text{ N} \cdot \text{m} \qquad \Diamond$$

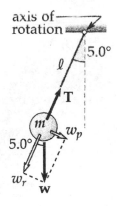

The second sketch illustrates an alternate method of computing the torque due to the force **w** about the axis of rotation. Note that **w** has **been replaced by its components**, w_r and w_p. The component w_r lies along the line connecting the bob and the rotation axis, while the component w_p is perpendicular to that line. Thus, the lever arm of w_r is zero (the line along its direction passes through the axis, or there is zero distance from this line to the axis), while the lever arm of w_p is the length of the string ℓ. The torque due to w can then be **computed as the sum of the torques due to its components** as follows:

$$\tau = (\text{torque of } w_r) + (\text{torque of } w_p)$$

$$\tau = w_r(0) + w_p(\ell) = 0 + (w\sin 5.0°)(\ell) = w\ell \sin 5.0°$$

This gives $\tau = 5.1 \text{ N} \cdot \text{m}$, the same as the other method. In many cases, this will be the easiest way to compute the torque due to a force. The reason for this is that the lever arms of the components are often easier to compute than the lever arm of the original force.

---

9. A cook holds a 2.00-kg carton of milk at arm's length (Fig. P8.9). What force $\mathbf{F}_B$ must be exerted by the biceps muscle? (Ignore the weight of the forearm.)

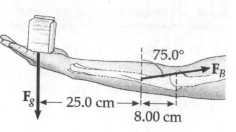

Solution For the system consisting of the forearm and the milk carton to be in equilibrium, it is necessary that the sum of the torques be zero about any axis we choose.

Figure P8.9

The upper arm exerts some unknown force on the forearm at the elbow. However, if we select a rotation axis that is perpendicular to the page and passes through the elbow, this unknown force will have zero lever arm and exert zero torque about our axis.

Resolve the force exerted by the biceps muscle into horizontal and vertical components:

$$(F_B)_x = F_B \sin 75.0° \qquad \text{and} \qquad (F_B)_y = F_B \cos 75.0°$$

The line along which the horizontal component acts passes through the chosen rotation axis, so that component exerts zero torque. Applying the second condition for equilibrium to the system (forearm and milk carton) yields

$$\Sigma\tau = +F_g(25.0\text{ cm} + 8.00\text{ cm}) - (F_B)_y(8.00\text{ cm}) = 0$$

or $\quad mg(25.0\text{ cm} + 8.00\text{ cm}) - (F_B \cos 75.0°)(8.00\text{ cm}) = 0$

The force exerted by the biceps muscle is then given by

$$F_B = \frac{\left[(2.00\text{ kg})(9.80\text{ m}/\text{s}^2)\right](33.0\text{ cm})}{(8.00\text{ cm})\cos 75.0°} = 312\text{ N}$$

21. A uniform semicircular sign 1.00 m in diameter and of weight w is supported by two wires as shown in Figure P8.21. What is the tension in each of the wires supporting the sign?

Solution Consider the free-body diagram of the sign shown at the right. Note that the center of gravity lies on the axis AA′ since the sign is symmetric about that axis. The sign is in equilibrium. Thus, the second condition of equilibrium states that $\Sigma\tau = 0$ for **any** rotation axis.

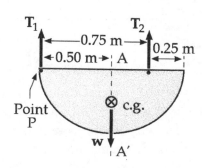

Figure P8.21

118

For convenience, choose a rotation axis that passes through point P perpendicular to the page. Note that this axis was chosen because it is observed that the lever arm of the unknown force T_1 (and hence the torque due to T_1) will be zero for this axis. This will considerably simplify the torque equation. Writing out the torque equation for this axis, taking counterclockwise torques as positive, gives:

$$T_1(0) - w(0.50 \text{ m}) + T_2(0.75 \text{ m}) = 0 \qquad \text{or} \qquad T_2 = \frac{(0.50 \text{ m})w}{(0.75 \text{ m})} = \frac{2}{3}w \qquad \Diamond$$

Then, applying the first condition of equilibrium, $\Sigma F_x = 0 \qquad$ and $\qquad \Sigma F_y = 0$

to the forces acting on the sign gives $\qquad +T_1 - w + T_2 = 0$

Therefore, $\qquad T_1 = w - \frac{2}{3}w = \frac{1}{3}w \qquad \Diamond$

---

25. An 8.00-m, 200-N uniform ladder rests against a smooth wall. The coefficient of static friction between the ladder and the ground is 0.600, and the ladder makes a 50.0° angle with the ground. How far up the ladder can an 800-N person climb before the ladder begins to slip?

Solution The sketch to the right gives the free-body diagram of the ladder after the 800-N person has gone a distance ℓ up the ladder. Since the ladder is uniform, the center of gravity is at its geometric center, or 4.00 m from either end.

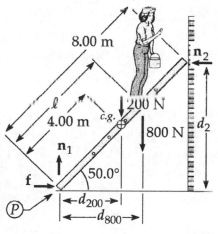

Gravity tends to cause the upper end of the ladder to slide down the wall with the lower end moving to the left along the ground. Thus, the friction force at the ground is toward the right to oppose this motion. Since the wall is smooth, there is no friction force at the upper end of the ladder.

Until slippage occurs, the ladder is in equilibrium, so $\Sigma F_x = 0$ and $\Sigma F_y = 0$. Further, when slippage is just about to occur, the static friction force will be at its maximum value: $f = (f_s)_{max}$.

$\sum F_y = 0 \qquad$ gives $\qquad +n_1 - 200 \text{ N} - 800 \text{ N} = 0 \qquad$ or $\qquad n_1 = 1000 \text{ N}$

$\qquad\qquad\qquad\qquad f = (f_s)_{max} = \mu_s n_1 \qquad\qquad$ or $\qquad f = (0.600)(1000 \text{ N}) = 600 \text{ N}$

$\sum F_x = 0 \qquad$ gives $\qquad +f - n_2 = 0 \qquad\qquad\qquad$ or $\qquad n_2 = f = 600 \text{ N}$

The second condition of equilibrium ($\Sigma \tau = 0$ for any rotation axis) must also be satisfied. Choosing a rotation axis perpendicular to the page through point P, the lever arms of n_1 and f are both zero since these forces pass through point P. The lever arms of the other forces are shown in the sketch and have values of $d_{200} = (4.00 \text{ m})\cos 50.0°$, $d_{800} = \ell \cos 50.0°$, and $d_2 = (8.00 \text{ m})\sin 50.0°$. The equation for the second condition of equilibrium (taking counter-clockwise torques as positive) is then:

$$\sum \tau = -(200 \text{ N})(4.00 \text{ m})\cos 50.0° - (800 \text{ N})\ell \cos 50.0° + (600 \text{ N})(8.00 \text{ m})\sin 50.0° = 0$$

Thus, $\ell (800 \text{ N})\cos 50.0° = -(800 \text{ N} \cdot \text{m})\cos 50.0° + (4800 \text{ N} \cdot \text{m})\sin 50.0°$

or $\quad \ell = \dfrac{-(800 \text{ N} \cdot \text{m})\cos 50.0° + (4800 \text{ N} \cdot \text{m})\sin 50.0°}{(800 \text{ N})\cos 50.0°} = 6.15 \text{ m}$

The person can go 6.15 m up the ladder (and no farther) before the ladder will begin to slip. ◊

---

29. Four objects are held in position at the corners of a rectangle by light rods as shown in Figure P8.29. Find the moment of inertia of the system about (a) the x axis, (b) the y axis, and (c) an axis through O and perpendicular to the page.

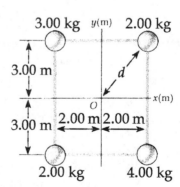

Figure P8.29

Solution

The moment of inertia of a body when that body is rotating about some axis is defined as $I \equiv \Sigma m_i r_i^2$ where r_i is the shortest distance from the mass to the rotation axis.

(a) If the body rotates about the x-axis, each of the four objects in Figure P8.29 will follow a circular path 3.00 m in radius. Thus, the moment of inertia of this body for rotation about the x-axis is:

$$I_x = (3.00 \text{ kg})(3.00 \text{ m})^2 + (2.00 \text{ kg})(3.00 \text{ m})^2 +$$

$$(4.00 \text{ kg})(3.00 \text{ m})^2 + (2.00 \text{ kg})(3.00 \text{ m})^2 = 99.0 \text{ kg} \cdot \text{m}^2 \qquad ◊$$

(b)   When this body rotates about the y-axis, each of the objects follow a circular path 2.00 m in radius. The moment of inertia of the body when rotating about the y-axis is therefore:

$$I_y = (3.00 \text{ kg})(2.00 \text{ m})^2 + (2.00 \text{ kg})(2.00 \text{ m})^2 +$$
$$(4.00 \text{ kg})(2.00 \text{ m})^2 + (2.00 \text{ kg})(2.00 \text{ m})^2 = 44.0 \text{ kg} \cdot \text{m}^2 \qquad \diamond$$

(c)   If the body rotates about an axis perpendicular to the page and passing through point O, each of the objects will follow a circular path of radius d, where

$$d = \sqrt{(2.00 \text{ m})^2 + (3.00 \text{ m})^2} = \sqrt{13} \text{ m}$$

The moment of inertia when the body rotates about this axis is then:

$$I_O = (3.00 \text{ kg})(\sqrt{13} \text{ m})^2 + (2.00 \text{ kg})(\sqrt{13} \text{ m})^2 +$$
$$(4.00 \text{ kg})(\sqrt{13} \text{ m})^2 + (2.00 \text{ kg})(\sqrt{13} \text{ m})^2 = 143 \text{ kg} \cdot \text{m}^2 \qquad \diamond$$

---

36.   A cylindrical 5.00-kg reel with a radius of 0.600 m and a frictionless axle, starts from rest and speeds up uniformly as a 3.00-kg bucket falls into a well, making a light rope unwind from the reel (Fig. P8.36). The bucket starts from rest and falls for 4.00 s. (a) What is the linear acceleration of the falling bucket? (b) How far does it drop? (c) What is the angular acceleration of the reel?

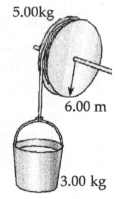

6.00 m

3.00 kg

Solution   Free-body diagrams of the reel and the bucket are given in the sketch to the right. When the system is released, the bucket will accelerate downward with an acceleration of magnitude a. The reel will rotate counter-clockwise with angular acceleration α. If the cord from the bucket does not slip on the reel, the magnitudes of these accelerations are related

Figure P8.36

as $a = R\alpha$ or $\alpha = a / R$ [1]

(a)   To determine the magnitudes of these accelerations, it is necessary to apply Newton's second law to each body separately.

For the rotational motion of the reel, Newton's second law takes the form $\Sigma \tau = I\alpha$, where I is the moment of inertia of the reel about a rotation axis perpendicular to the page and through its center.

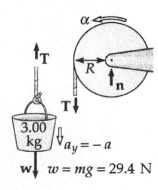

If the reel is a solid cylindrical body,

$$I = \tfrac{1}{2}m_r R^2 = \tfrac{1}{2}(5.00 \text{ kg})(0.600 \text{ m})^2 = 0.900 \text{ kg} \cdot \text{m}^2$$

Then, taking counterclockwise rotations as positive,

$$\sum \tau = +TR = I\alpha: \qquad T = \left(\frac{I}{R}\right)\alpha = \left(\frac{0.900 \text{ kg} \cdot \text{m}^2}{0.600 \text{ m}}\right)\alpha = (1.50 \text{ kg} \cdot \text{m})\alpha$$

Using Equation [1] to substitute for α in this result gives

$$T = (1.50 \text{ kg} \cdot \text{m})\left(\frac{a}{0.600 \text{ m}}\right) = (2.50 \text{ kg})a \qquad\qquad [2]$$

Taking upward as positive and applying Newton's second law to the bucket,

$$\sum F_y = ma_y: \qquad\qquad +T - 29.4 \text{ N} = (3.00 \text{ kg})(-a)$$

or

$$T = 29.4 \text{ N} - (3.00 \text{ kg})a \qquad\qquad [3]$$

Combining Equations [2] and [3] gives the **downward** acceleration of the bucket

as

$$(2.50 \text{ kg})a = 29.4 \text{ N} - (3.00 \text{ kg})a$$

or

$$a = \frac{29.4 \text{ N}}{5.50 \text{ kg}} = 5.35 \text{ m} / \text{s}^2 \qquad\qquad \Diamond$$

(b) The vertical displacement of the bucket at time t is $y = v_{iy}t + \tfrac{1}{2}a_y t^2$. The bucket starts from rest, so $v_{iy} = 0$. Still taking upward as positive, the vertical acceleration of the bucket is

$$a_y = -a = -5.35 \text{ m} / \text{s}^2$$

Thus, at $t = 4.00$ s, $y = 0 + \tfrac{1}{2}\left(-5.35 \text{ m} / \text{s}^2\right)(4.00 \text{ s})^2 = -42.8 \text{ m}$

The displacement of the bucket after 4.00 s is   42.8 m **downward** $\Diamond$

(c) From Equation [1], the angular acceleration of the cylindrical reel is

$$\alpha = \frac{a}{R} = \frac{5.35 \text{ m} / \text{s}^2}{0.600 \text{ m}} \quad \text{or} \quad \alpha = 8.91 \text{ rad} / \text{s}^2 \qquad \textbf{(counterclockwise)} \qquad \Diamond$$

39. A 10.0-kg cylinder rolls without slipping on a rough surface. At the instant when its center of gravity has a speed of 10.0 m/s, determine (a) the translational kinetic energy of its center of gravity, (b) the rotational kinetic energy about its center of gravity, and (c) its total kinetic energy.

Solution

(a) The translational kinetic energy of the cylinder is given by $KE_t = \frac{1}{2}mv^2$, where v is the translational speed of the center of gravity.

Thus, $KE_t = \frac{1}{2}(10.0 \text{ kg})(10.0 \text{ m / s})^2$ or $KE_t = 500 \text{ J}$ ◊

(b) The rotational kinetic energy of the cylinder is $KE_r = \frac{1}{2}I\omega^2$

where I is the moment of inertia about the axis through its center of gravity, and ω is the angular speed about this axis. Assuming a uniform, solid cylinder of radius R,

$$I = \frac{1}{2}mR^2$$

the rotational kinetic energy is $KE_r = \frac{1}{2}\left(\frac{1}{2}mR^2\right)\omega^2 = \frac{1}{4}m(R\omega)^2$

The product $R\omega$ is the same as the tangential speed of a point on the rim of the cylinder,

$$v_t = R\omega \qquad\qquad \text{so} \qquad KE_r = \frac{1}{4}mv_t^2$$

If the wheel rolls without slipping, the tangential speed of a point on the rim is the same as the translational speed of the center of gravity. To understand why this is true, imagine yourself to be at the center of the wheel and moving to your left at speed v. Looking down, you would see point A at the wheel's rim moving to your right with the tangential speed $v_t = R\omega$. You would also see point B on the ground moving to your right at the speed v, the speed of the center of gravity (and you) relative to the ground.

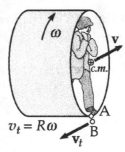

Now, if the rim of the wheel is not slipping against the ground, the points A and B (in contact with each other) must move at the same speed. Thus, it is necessary that $v = v_t = R\omega$ if slipping does not occur. The rotational kinetic energy of the wheel is therefore,

$$KE_r = \frac{1}{4}mv_t^2 = \frac{1}{4}mv^2 = \frac{1}{4}(10.0 \text{ kg})(10.0 \text{ m / s})^2 = 250 \text{ J} \qquad ◊$$

(c) The total kinetic energy of the rolling wheel is then:

$$KE_{total} = KE_t + KE_r = 500 \text{ J} + 250 \text{ J} = 750 \text{ J} \qquad ◊$$

43. The top in Figure P8.43 has a moment of inertia of 4.00×10^{-4} kg·m^2 and is initially at rest. It is free to rotate about a stationary axis, AA'. A string, wrapped around a peg along the axis of the top, is pulled in such a manner as to maintain a constant tension of 5.57 N in the string. If the string does not slip while wound around the peg, what is the angular speed of the top after 80.0 cm of string has been pulled off the peg? (**Hint:** Consider the work done.)

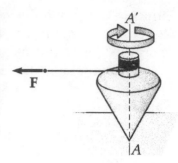

Figure P8.43

Solution

The agent pulling on the end of the string does work on the system consisting of the string and top. When the end of the string has been moved distance s, the work done on this system is

$$W_{net} = Fs\cos 0° \quad \text{where } F \text{ is the constant tension maintained in the string.}$$

Neglecting the mass of the string, the only kinetic energy of the system is the rotational kinetic energy of the top. The work-kinetic energy theorem, $W_{net} = KE_f - KE_i$, then becomes

$$Fs\cos 0° = \frac{1}{2}I_{top}\omega_f^2 - \frac{1}{2}I_{top}\omega_i^2$$

If the top starts from rest and the tension maintained in the string is $F = 5.57$ N, the angular speed of the top after the end of the string has been moved 80.0 cm is

$$\omega_f = \sqrt{\frac{2Fs\cos 0°}{I_{top}}} = \sqrt{\frac{2(5.57 \text{ N})(0.800 \text{ m})(1)}{4.00 \times 10^{-4} \text{ kg·m}^2}} = 149 \text{ rad / s} \qquad \lozenge$$

---

51. The puck in Figure P8.51 has a mass of 0.120 kg. Its original distance from the center of rotation is 40.0 cm, and the puck is moving with a speed of 80.0 cm/s. The string is pulled downward 15.0 cm through the hole in the frictionless table. Determine the work done on the puck.

(**Hint:** Consider the change of kinetic energy of the puck.)

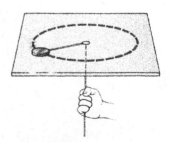

Figure P8.51

Solution   The puck originally moves in a circular path with a radius of

$$r_i = 40.0 \text{ cm} = 0.400 \text{ m}$$

Pulling the string downward 15.0 cm decreases the radius of the puck's path to

$$r_f = 0.400 \text{ m} - 0.150 \text{ m} = 0.250 \text{ m}.$$

The tension in the string is directed toward the center of the puck's circular path, and hence, toward the rotation axis.   It exerts zero torque about the rotation axis.   The weight of the puck and the normal force acting on the puck are both parallel to the rotation axis and therefore exert zero torque about that axis.   The total torque acting about the rotation axis of the puck is then zero, and the angular momentum of the puck remains constant:

$$I_f \omega_f = I_i \omega_i$$

The moments of inertia about the rotation axis before and after the string is pulled are·

$$I_i = m r_i^2 = (0.120 \text{ kg})(0.400 \text{ m})^2 = 1.92 \times 10^{-2} \text{ kg} \cdot \text{m}^2$$

and $$I_f = m r_f^2 = (0.120 \text{ kg})(0.250 \text{ m})^2 = 7.50 \times 10^{-3} \text{ kg} \cdot \text{m}^2$$

The initial angular speed of the puck is

$$\omega_i = \frac{v_i}{r_i} = \frac{80.0 \text{ cm/s}}{40.0 \text{ cm}} = 2.00 \text{ rad/s}$$

and the final angular speed is

$$\omega_f = \left(\frac{I_i}{I_f}\right)\omega_i = \left(\frac{1.92 \times 10^{-2} \text{ kg} \cdot \text{m}^2}{7.50 \times 10^{-3} \text{ kg} \cdot \text{m}^2}\right)(2.00 \text{ rad/s}) = 5.12 \text{ rad/s}$$

The kinetic energy of the puck before and after the string is pulled is then:

$$KE_i = \frac{1}{2} I_i \omega_i^2 = \frac{1}{2}\left(1.92 \times 10^{-2} \text{ kg} \cdot \text{m}^2\right)(2.00 \text{ rad/s})^2 = 3.84 \times 10^{-2} \text{ J}$$

$$KE_f = \frac{1}{2} I_f \omega_f^2 = \frac{1}{2}\left(7.50 \times 10^{-3} \text{ kg} \cdot \text{m}^2\right)(5.12 \text{ rad/s})^2 = 9.83 \times 10^{-2} \text{ J}$$

Thus, work done on the puck is given by the work-kinetic energy theorem as

$$W_{net} = KE_f - KE_i = 9.83 \times 10^{-2} \text{ J} - 3.84 \times 10^{-2} \text{ J} = 5.99 \times 10^{-2} \text{ J}$$ ◊

63. A solid 2.0-kg ball of radius 0.50 m starts at a height of 3.0 m above the surface of the Earth and **rolls** down a 20° slope. A solid disk and a ring start at the same time and the same height. The ring and disk each have the same mass and radius as the ball. Which of the three wins the race to the bottom if all roll without slipping?

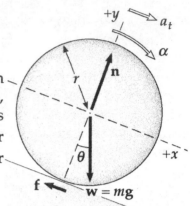

Solution

Consider the free-body diagram of an object rolling down the incline. The object could be either a solid sphere (ball), a solid cylinder (disk), or a hoop (ring). If the object rolls without slipping, then $a_t = r\alpha$ where a_t is the linear acceleration of the center of gravity and α is the angular acceleration about the rotation axis.

From $\quad \Sigma F_x = ma_x \quad$ we obtain $\quad mg\sin\theta - f = ma_t$ [1]

Now, consider an axis perpendicular to the page and passing through the center of the object.

$$\sum \tau = I\alpha \qquad \text{becomes} \qquad f \cdot r = I\alpha = I\left(\frac{a_t}{r}\right) \qquad \text{or} \qquad f = \left(\frac{I}{r^2}\right)a_t$$

Substitute this result into Equation [1] and simplify to obtain $\quad a_t = \dfrac{g\sin\theta}{\left(1 + I/mr^2\right)}$

as the linear acceleration of the center of gravity of the object.

For a solid sphere, $\qquad\qquad I = \frac{2}{5}mr^2 \qquad\qquad$ so $\qquad a_{\text{sphere}} = \dfrac{g\sin\theta}{1.4}$

For a solid cylinder, $\qquad\qquad I = \frac{1}{2}mr^2 \qquad\qquad$ and $\qquad a_{\text{cylinder}} = \dfrac{g\sin\theta}{1.5}$

Finally, for a hoop, $\qquad\qquad I = mr^2 \qquad\qquad$ so $\qquad a_{\text{ring}} = \frac{1}{2}g\sin\theta$

Thus, we find $\qquad\qquad\qquad a_{\text{sphere}} > a_{\text{cylinder}} > a_{\text{ring}}$

so the ball wins the race, the disk comes in second, and the ring is last. ◊

Note that each of the calculated accelerations is independent of the both the mass and the radius of the object. Thus, the three objects need not be of the same mass or size. Only the shapes and the distribution of the mass within the objects affect the outcome of the race.

68. Two astronauts (Fig. P8.68), each having a mass of 75.0 kg, are connected by a 10.0-m rope of negligible mass. They are isolated in space, moving in circles around the point halfway between them at speeds of 5.00 m/s. Treating the astronauts as particles, calculate (a) the magnitude of the angular momentum and (b) the rotational energy of the system. By pulling on the rope, the astronauts shorten the distance between them to 5.00 m. (c) What is the new angular momentum of the system? (d) What are their new speeds? (e) What is the new rotational energy of the system? (f) How much work is done by the astronauts in shortening the rope?

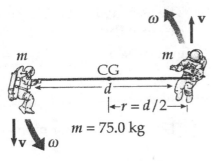

Figure P8.68

$m = 75.0$ kg

Solution

(a) Initially, the astronauts are moving in a circular path

of a radius $\qquad r_i = (10.0 \text{ m})/2 = 5.00$ m

with linear speeds of $\qquad v_i = 5.00$ m/s

and angular speeds of $\qquad \omega_i = \dfrac{v}{r} = \dfrac{5.00 \text{ m/s}}{5.00 \text{ m}} = 1.00$ rad/s

The initial moment of inertia and initial angular momentum of this system are:

$$I_i = 2I_{\substack{\text{individual} \\ \text{astronaut}}} = 2\left(m r_i^2\right) = 2(75.0 \text{ kg})(5.00 \text{ m})^2 = 3.75 \times 10^3 \text{ kg} \cdot \text{m}^2$$

and $\quad L_i = I_i \omega_i = \left(3.75 \times 10^3 \text{ kg} \cdot \text{m}^2\right)(1.00 \text{ rad/s}) = 3.75 \times 10^3 \text{ kg} \cdot \text{m}^2/\text{s}$ ◊

(b) The initial rotational kinetic energy of this system is $\quad KE_i = \frac{1}{2} I_i \omega_i^2$

$$KE_i = \frac{1}{2}\left(3.75 \times 10^3 \text{ kg} \cdot \text{m}^2\right)(1.00 \text{ rad/s})^2 = 1.88 \times 10^3 \text{ J} = 1.88 \text{ kJ}$$ ◊

(c) Since the astronauts are isolated in space, no external forces exert torques about the rotation axis. Thus, the angular momentum of the system is constant.

That is, $\qquad L_f = L_i = 3.75 \times 10^3 \text{ kg} \cdot \text{m}^2/\text{s}$ ◊

(d) The astronauts change the moment of inertia of the system when they change the distance they are from the center of gravity. The new distance from the center of gravity is $r_f = (5.00 \text{ m})/2 = 2.50 \text{ m}$, and the new moment of inertia is

$$I_f = 2(75.0 \text{ kg})(2.50 \text{ m})^2 = 938 \text{ kg} \cdot \text{m}^2$$

Then, since angular momentum is conserved, $I_f \omega_f = I_i \omega_i$

and
$$\omega_f = \frac{I_i \omega_i}{I_f} = \left(\frac{3.75 \times 10^3 \text{ kg} \cdot \text{m}^2/\text{s}}{938 \text{ kg} \cdot \text{m}^2} \right)(1.00 \text{ rad}/\text{s}) = 4.00 \text{ rad}/\text{s}$$

The new linear speed is

$$v_f = r_f \omega_f = (2.50 \text{ m})(4.00 \text{ rad}/\text{s}) = 10.0 \text{ m}/\text{s} \qquad \Diamond$$

(e) The new rotational kinetic energy of this system is $KE_f = \frac{1}{2} I_f \omega_f^2$

or
$$KE_f = \frac{1}{2}(938 \text{ kg} \cdot \text{m}^2)(4.00 \text{ rad/s})^2 = 7.50 \times 10^3 \text{ J} = 7.50 \text{ kJ}$$

(f) The work done is $W_{\text{net}} = KE_f - KE_i = 7.50 \text{ kJ} - 1.88 \text{ kJ} = 5.62 \text{ kJ}$ \qquad \Diamond

---

73. A uniform, solid cylinder of mass M and radius R rotates on a frictionless horizontal axle (Fig. P8.73). Two objects with equal masses hang from light cords wrapped around the cylinder. If the system is released from rest, find (a) the tension in each cord and (b) the acceleration of each object after the objects have descended a distance h.

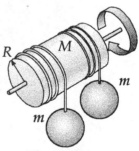

Figure P8.73

Solution

In the free-body diagrams at the right, the two falling masses have been combined into one object of mass $2m$. The total downward force these masses exert on the cylinder is $2T$ where T is the tension in either cord.

If the cords do not slip on the cylinder, the linear acceleration of each falling object is related to the angular acceleration of the cylinder by $a_t = R\alpha$.

Choose an axis perpendicular to the page and passing through the center of the cylinder.

Then, applying $\Sigma\tau = I\alpha$ to the cylinder gives

$$(2T)\cdot R = \left(\tfrac{1}{2}MR^2\right)\alpha = \left(\tfrac{1}{2}MR^2\right)\left(\frac{a_t}{R}\right) \qquad \text{or} \qquad T = \tfrac{1}{4}Ma_t \qquad\qquad \text{[1]}$$

Now apply $\Sigma F_y = ma_y$ to the falling objects to obtain

$$(2m)g - 2T = (2m)a_t \qquad\qquad \text{or} \qquad a_t = g - \frac{T}{m} \qquad\qquad \text{[2]}$$

(a)   Substituting Equation [2] into Equation [1]

yields
$$T = \frac{Mg}{4} - \left(\frac{M}{4m}\right)T$$

which reduces to
$$T = \frac{Mmg}{M + 4m} \qquad\qquad \Diamond$$

(b)   From Equation [2], the acceleration of each falling object is found to be

$$a_t = g - \frac{1}{m}\left(\frac{Mmg}{M + 4m}\right) = g - \frac{Mg}{M + 4m} - \frac{4mg}{M + 4m} \qquad\qquad \Diamond$$

Note that the results are independent of the distance fallen, h.

Chapter 9
SOLIDS AND FLUIDS

NOTES ON SELECTED CHAPTER SECTIONS

9.1   States of Matter

Matter is generally classified as being in one of three states: solid, liquid, or gas.

In a **crystalline solid**, the atoms are arranged in an ordered periodic structure; while in an **amorphous solid** (i.e. glass), the atoms are present in a disordered fashion.

In the **liquid state**, thermal agitation is greater than in the solid state, the molecular forces are weaker, and molecules wander throughout the liquid in a random fashion.

The molecules of a **gas** are in constant random motion and exert weak forces on each other. The distances separating molecules are large compared to the dimensions of the molecules.

9.2   The Deformation of Solids

The elastic properties of solids are described in terms of stress and strain. Stress is a quantity that is related to the force causing a deformation; strain is a measure of the degree of deformation. It is found that, for sufficiently small stresses, stress is proportional to strain, and the constant of proportionality depends on the material being deformed and on the nature of the deformation. We call this proportionality constant the elastic modulus.

We shall consider three types of deformation and define an elastic modulus for each:

1. **Young's modulus**, which measures the resistance of a solid to a change in its length.

2. **Shear modulus**, which measures the resistance to displacement of the planes of a solid sliding past each other.

3. **Bulk modulus**, which measures the resistance that solids or liquids offer to changes in their volume.

9.3   Density and Pressure

The **density**, ρ, of a substance of uniform composition is defined as its **mass per unit volume** and has units of kilograms per cubic meter (kg/m^3) in the SI system.

The **specific gravity** of a substance is a dimensionless quantity which is the ratio of the density of the substance to the density of water.

The **pressure**, P, in a fluid is the force per unit area that the fluid exerts on an object immersed in the fluid.

9.4   Variation of Pressure with Depth

In a **fluid at rest**, all points at the same depth are at the same pressure. **Pascal's law** states that a change in pressure applied to an **enclosed** fluid is transmitted undiminished to every point in the fluid and the walls of the containing vessel. **The pressure, P, at a depth of h below the surface of a liquid open to the atmosphere is greater than atmospheric pressure by an amount ρgh.**

9.5   Pressure Measurements

The **absolute pressure** of a fluid is the sum of the **gauge pressure** and **atmospheric pressure**. The SI unit of pressure is the Pascal (Pa). Note that $1\,Pa \equiv 1\,N/m^2$.

9.6   Buoyant Forces and Archimedes's Principle

Any object partially or completely submerged in a fluid experiences a buoyant force equal in magnitude to the weight of the fluid displaced by the object and acting vertically upward through the point which was the center of gravity of the displaced fluid.

9.7   Fluids in Motion

Many features of fluid motion can be understood by considering the behavior of an ideal fluid, which satisfies the following conditions:

1. **The fluid is nonviscous**; that is, there is no internal friction force between adjacent fluid layers.

2. **The fluid is incompressible**, which means that its density is constant.

3. **The fluid motion is steady**, meaning that the velocity, density, and pressure at each point in the fluid do not change in time.

4. **The fluid moves without turbulence**. This implies that each element of the fluid has zero angular velocity about its center; that is, there can be no eddy currents present in the moving fluid.

Fluids which have the "ideal" properties stated above obey two important equations:

1. The **equation of continuity** states that the flow rate through a pipe is constant (i.e. the product of the cross-sectional area of the pipe and the speed of the fluid is constant).

2. **Bernoulli's equation** states that the sum of the pressure (P), kinetic energy per unit volume ($\rho v^2 / 2$), and the potential energy per unit volume ($\rho g h$) has a constant value at all points along a streamline.

9.9   Surface Tension, Capillary Action, and Viscous Fluid Flow

The concept of surface tension can be thought of as the energy content of the fluid at its surface per unit surface area. In general, **any equilibrium configuration of an object is one in which the energy is minimum.** For a given volume, the spherical shape is the one that has the smallest surface area; therefore, a drop of water takes on a spherical shape. The surface tension of liquids decreases with increasing temperature.

Forces between like molecules, such as the forces between water molecules, are called cohesive forces and forces between unlike molecules, such as those of glass on water are **adhesive forces**. If a capillary tube is inserted into a fluid for which adhesive forces dominate over cohesive forces, the surrounding liquid will rise into the tube. If a capillary tube is inserted into a liquid in which cohesive forces dominate over adhesive forces, the level of the liquid in the capillary tube will be below the surface of the surrounding fluid.

Viscosity refers to the internal friction of a fluid. At sufficiently high velocities, fluid flow changes from simple streamline flow to turbulent flow. The onset of turbulence in a tube is determined by a factor called the **Reynolds number**, which is a function of the density of the fluid, the average speed of the fluid along the direction of flow, the diameter of the tube, and the viscosity of the fluid.

9.10 Transport Phenomena

The two fundamental processes involved in fluid transport resulting from concentration differences are called **diffusion** and **osmosis**. In a **diffusion** process, molecules move from a region where their concentration is high to a region where their concentration is lower. Diffusion occurs readily in air; the process also occurs in liquids and, to a lesser extent, in solids. **Osmosis** is defined as the movement of water from a region where its concentration is high, across a **selectively permeable membrane**, into a region where its concentration is lower. Osmosis is often described simply as the diffusion of water across a membrane.

EQUATIONS AND CONCEPTS

A body of matter can be deformed (experience change in size or shape) by the application of external forces. Stress is a quantity which is proportional to the force which causes the deformation; strain is a measure of the degree of deformation. The elastic modulus is a general characterization of the deformation. Specific deformations are characterized by specific moduli.

$$\text{Elastic modulus} = \frac{\text{stress}}{\text{strain}} \qquad (9.1)$$

The SI units of pressure are newtons per square meter, or Pascals (Pa).

$$1\,\text{Pa} \equiv 1\,\text{N}/\text{m}^2 \qquad (9.2)$$

Young's modulus is a measure of the resistance of a body to elongation. It is defined as the ratio of tensile stress to tensile strain.

$$Y \equiv \frac{\text{tensile stress}}{\text{tensile strain}} = \frac{F / A}{\Delta L / L_0} = \frac{FL_0}{A\Delta L} \quad (9.3)$$

Within a limited range of values, the graph of stress vs. strain for a given substance will be a straight line. When the stress exceeds the elastic limit (at the yield point), the stress-strain curve will no longer be linear.

Comment on experimental observations

Consider a rectangular block of height h and top and bottom surfaces each of area A. A force **F** applied parallel to the top surface, with the bottom surface fixed, will cause the top surface to move forward a distance Δx. **The ratio F/A is the shear stress and the ratio $\Delta x/h$ is the shear strain.**

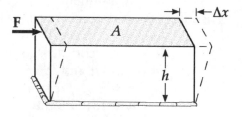

The **Shear modulus** is a measure of the deformation which occurs when a force is applied along a direction parallel to one surface of a body.

$$S \equiv \frac{\text{shear stress}}{\text{shear strain}} = \frac{F / A}{\Delta x / h} \quad (9.4)$$

The **Bulk modulus** characterizes the response of a body to uniform pressure (or squeezing) on all sides. Note that when ΔP is positive (increase in pressure), the ratio $\Delta V/V$ will be negative (decrease in volume) and vice versa. Therefore, the negative sign in the equation ensures that B will always be positive.

$$B \equiv \frac{\text{volume stress}}{\text{volume strain}} = -\frac{\Delta P}{\Delta V / V} \quad (9.5)$$

The **density** of a homogeneous substance is defined as its ratio of mass per unit volume. The value of density is characteristic of a particular type of material and independent of the total quantity of material in the sample.

$$\rho \equiv \frac{M}{V} \qquad (9.6)$$

The SI units of density are kg per cubic meter.

$$1 \, g \, / \, cm^3 = 1\,000 \, kg \, / \, m^3$$

The (average) pressure of a fluid is defined as the normal force per unit area acting on a surface immersed in the fluid. Pressure has units of pascals (newtons per square meter).

$$P \equiv \frac{F}{A} \qquad (9.7)$$

Atmospheric pressure is often expressed in other units:

Conversion of units

atmospheres

$$1 \, atm = 1.013 \times 10^5 \, Pa$$

mm of mercury (Torr)

$$1 \, Torr = 133.3 \, Pa$$

pounds per square inch

$$1 \, lb \, / \, in^2 = 6.895 \times 10^3 \, Pa$$

The absolute pressure, P, at a depth, h, below the surface of a liquid which is open to the atmosphere is greater than atmospheric pressure, P_0, by an amount which depends on the depth below the surface. The pressure at a depth h below the surface of a liquid does not depend on the shape of the container.

$$P = P_0 + \rho g h \qquad (9.10)$$

$$P_0 = 1.013 \times 10^5 \, Pa$$

The quantity $\rho g h$ is called the gauge pressure and P is the absolute pressure. Therefore,

$$\frac{\text{absolute}}{\text{pressure}} = \frac{\text{atmospheric}}{\text{pressure}} + \frac{\text{gauge}}{\text{pressure}}$$

Pascal's law states that pressure applied to an **enclosed fluid** (liquid or gas) is transmitted undiminished to every point within the fluid and over the walls of the vessel which contain the fluid.

Archimedes's principle states that when an object is partially or fully submerged in a fluid, the fluid exerts an upward buoyant force, B, which depends on the fluid density and the displaced volume. **The buoyant force equals the weight of the displaced fluid.**

Fluid dynamics, the treatment of fluids in motion, is greatly simplified under the assumption that the fluid is ideal with the following characteristics:

(a) nonviscous — internal friction between adjacent fluid layers is negligible

(b) incompressible — the density throughout the fluid is constant.

(c) steady — the velocity, density, and pressure at each point in the fluid are constant in time

(d) irrotational (without turbulence) — there are no eddy currents within the fluid (each element of the fluid has zero angular velocity about its center)

Comments on fluid pressure.

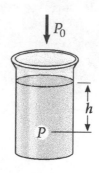

$$B = \rho_{\text{fluid}} V g = w_{\text{fluid}} \qquad (9.11)$$

For a submerged object:
$\quad V = $ Volume of object

For a floating object:
$\quad V = $ Volume displaced

Comments on ideal fluids.

For an **incompressible fluid** (ρ = constant), the equation of continuity can be written as Equation 9.14. The product is called the flow rate. Therefore, the **flow rate** at any point along a pipe carrying an incompressible fluid is constant.

$$A_1v_1 = A_2v_2 \qquad (9.14)$$

$$\text{(for incompressible fluid)}$$

Bernoulli's equation is the most fundamental law in fluid mechanics. It is a statement of the law of conservation of mechanical energy as applied to a fluid. Bernoulli's equation states that the sum of **pressure, kinetic energy per unit volume, and potential energy per unit volume remains constant** along a streamline of an ideal fluid.

$$P + \frac{1}{2}\rho v^2 + \rho g y = \text{constant} \qquad (9.16)$$

The **surface tension,** γ, in a film of liquid is defined as the ratio of the magnitude of the force along the surface force, **F**, to the length, L, along which the force acts. Note: If **F** is the surface tension force required to lift a small wire from the surface of a liquid, the value L in Equation 9.18 must be twice the length of the wire because the force of surface tension acts along both sides of the wire as it is being lifted from the liquid.

$$\gamma \equiv \frac{F}{L} \qquad (9.18)$$

In a capillary tube, the angle ϕ between the solid surface and a line drawn tangent to the liquid at the surface is called the **angle of contact**. Note that ϕ is less than 90° for any substance in which adhesive forces are stronger than cohesive forces. In such a case, the liquid will rise to a height h in the tube. If cohesive forces dominate, the liquid in the tube will be depressed below the surface of the surrounding liquid by a distance h.

$$h = \frac{2\gamma}{\rho g r}\cos\phi \qquad (9.21)$$

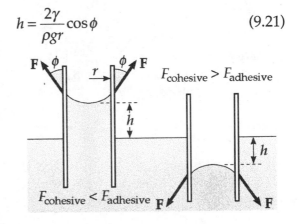

137

Consider a layer of liquid of thickness d and area A between two solid surfaces. The **coefficient of viscosity**, η (the lowercase Greek letter **eta**), for the fluid is defined in terms of the force, F, required to move one of the solid surfaces with a velocity v. The coefficient of viscosity may also be thought of as the ratio of the shearing stress to the rate of change of the shear strain. The SI units of viscosity are $N \cdot s/m^2$.

$$\eta \equiv \frac{Fd}{Av}$$

$$1 \text{ poise} = 10^{-1} \text{ N} \cdot \text{s} / \text{m}^2 \qquad (9.23)$$

The onset of turbulence in a tube is determined by a dimensionless factor called the **Reynolds number**, which is a function of the density of the fluid, the average speed of the fluid along the direction of flow, the diameter of the tube, and the viscosity of the fluid. In the region between $2\,000$ and $3\,000$, the flow is unstable, meaning that any small disturbance will cause its motion to change from streamline to turbulent flow.

$$RN = \frac{\rho v d}{\eta} \qquad (9.25)$$

Streamline flow: $\qquad RN < \; \sim 2\,000$

Unstable flow: $\qquad 2\,000 < RN < 3\,000$

Turbulent flow: $\qquad RN > 3\,000$

The basic equation for diffusion is Fick's law. The **diffusion rate**, a measure of the mass being transported per unit time, is proportional to the cross-sectional area A and to the change in concentration per unit distance, $(C_2 - C_1)/L$, which is called the concentration gradient. The proportionality constant D is called the **diffusion coefficient** and has units of square meters per second.

$$\begin{array}{c} \text{Diffusion} \\ \text{Rate} \end{array} = \frac{\Delta M}{\Delta t} = DA\left(\frac{C_2 - C_1}{L}\right) \qquad (9.26)$$

The resistive force on a small spherical object falling with speed v through as viscous fluid is given by Stokes' law.

$$F_r = 6\pi\eta r v \qquad (9.27)$$

138

As a sphere falls through a viscous medium, three forces act on it: the force of frictional resistance, the buoyant force of the fluid, and the weight of the sphere. When the net upward force balances the downward weight force, the sphere reaches terminal speed.

$$v_t = \frac{2r^2 g}{9\eta}\left(\rho - \rho_f\right) \qquad (9.28)$$

In a centrifuge those particles having the greatest mass will have the largest terminal speed. Therefore, the most massive particles will settle out of the mixture. The factor of k is a coefficient of frictional resistance which must be determined experimentally.

$$v_t = \frac{m\omega^2 r}{k}\left(1 - \frac{\rho_f}{\rho}\right) \qquad (9.31)$$

REVIEW CHECKLIST

▷ Describe the three types of deformations that can occur in a solid, and define the elastic modulus that is used to characterize each: (Young's modulus, Shear modulus, and Bulk modulus).

▷ Understand the concept of pressure at a point in a fluid, and the variation of pressure with depth. Understand the relationships among absolute, gauge, and atmospheric pressure values; and know the several different units commonly used to express pressure.

▷ Understand the origin of buoyant forces; and state and explain Archimedes's principle.

▷ State and understand the physical significance of the equation of continuity (constant flow rate) and Bernoulli's equation for fluid flow (relating flow velocity, pressure, and pipe elevation).

SOLUTIONS TO SELECTED END-OF-CHAPTER PROBLEMS

7.   Bone has a Young's modulus of about 18×10^9 Pa. Under compression, it can withstand a stress of about 160×10^6 Pa before breaking. Assume that a femur (thigh bone) is 0.50 m long and calculate the amount of compression this bone can withstand before breaking.

Solution

Young's Modulus for a material is defined as $Y = \text{stress}/\text{strain}$

where the tensile strain is the ratio of the change in length, ΔL to the original length, L_0. Thus, we can calculate the compression the material experiences as

$$Y = \frac{\text{stress}}{\Delta L / L_0}$$

or

$$\Delta L = \frac{L_0(\text{stress})}{Y}$$

The original length of the femur is $L_0 = 0.50$ m

and the femur breaks when the stress exceeds 160×10^6 Pa

Therefore, the maximum compression the femur can withstand is:

$$\Delta L = \frac{(0.50 \text{ m})(160 \times 10^6 \text{ Pa})}{18 \times 10^9 \text{ Pa}} = 4.4 \times 10^{-3} \text{ m} = 4.4 \text{ mm} \qquad \Diamond$$

13. A 50.0-kg ballet dancer stands on her toes during a performance with 4 in.^{2} $(26.0$ cm$^2)$ in contact with the floor. What is the pressure exerted by the floor over the area of contact (a) if the dancer is stationary, and (b) if the dancer is leaping upwards with an acceleration of 4.00 m $/$ s^2?

Solution

Two forces, both vertical, act on the body of the dancer as indicated in the sketch. The force F is the normal force exerted, by the floor, on the dancer's toe.

From Newton's second law,

$$\Sigma F_y = +F - w = m\,a_y$$

or $\qquad F = w + m\,a_y = m\left(g + a_y\right)$

(a) If the dancer is stationary, $a_y = 0$ and the normal force is

$$F = m\left(g + a_y\right) = \left(50.0 \text{ kg}\right)\left(9.80 \text{ m } / \text{ s}^2 + 0\right) = 490 \text{ N}$$

The pressure is then

$$P = \frac{F}{A} = \frac{490 \text{ N}}{26.0 \text{ cm}^2}\left[\frac{100 \text{ cm}}{1.00 \text{ m}}\right]^2 = 1.88 \times 10^5 \text{ N/m}^2 = 1.88 \times 10^5 \text{ Pa} \qquad \Diamond$$

(b) When $a_y = +4.00$ m/s^2

the force exerted by the floor is

$$F = \left(50.0 \text{ kg}\right)\left(9.80 \text{ m/s}^2 + 4.00 \text{ m/s}^2\right) = 690 \text{ N}$$

and the pressure is:

$$P = \frac{F}{A} = \frac{690 \text{ N}}{26.0 \text{ cm}^2}\left[\frac{100 \text{ cm}}{1.00 \text{ m}}\right]^2 = 2.65 \times 10^5 \text{ N/m}^2 = 2.65 \times 10^5 \text{ Pa} \qquad \Diamond$$

23. A container is filled to a depth of 20.0 cm with water. On top of the water floats a 30.0-cm-thick layer of oil with specific gravity 0.700. What is the absolute pressure at the bottom of the container?

Solution

If P_0 is the known absolute pressure at some chosen reference level, the absolute pressure at a depth h, in a fluid of density ρ, below that reference level is $P = P_0 + \rho g h$ where g is the acceleration due to gravity. In this problem, there are multiple fluids involved and it may appear ambiguous as to what density (i.e., density of air, oil or water) should be used in this equation for the pressure.

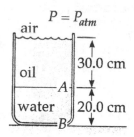

The answer to this dilemma is to deal with one fluid at a time. To find the absolute pressure at the bottom of the water layer (i.e., at the bottom of the container), first compute the absolute pressure at the boundary between the oil and water.

Atmospheric pressure exists at the upper surface of the oil (at the air-oil boundary). Choosing this boundary as the reference level, the absolute pressure at point A on the oil-water boundary is:

$$P_A = P_{atm} + \rho_{oil}\, g\, h_{oil}$$

The oil's specific gravity is $\quad (s.\,g.)_{oil} = \dfrac{\rho_{oil}}{\rho_{water}} = 0.700$

Thus, $\quad \rho_{oil} = (0.700)\rho_{water} = (0.700)(1000 \text{ kg}/m^3) = 700 \text{ kg}/m^3$

and, $\quad P_A = 1.013 \times 10^5 \text{ Pa} + (700 \text{ kg}/m^3)(9.80 \text{ m}/s^2)(0.300 \text{ m}) = 1.03 \times 10^5 \text{ Pa}$

Now that the pressure at point A is known, the oil-water boundary can be chosen as the new reference level (so $P_0 = P_A = 1.03 \times 10^5 \text{ Pa}$) and the absolute pressure at point B on the bottom of the container can be computed as

$$P_B = P_A + \rho_{water}\, g\, h_{water}$$

This yields $\quad P_B = 1.03 \times 10^5 \text{ Pa} + (1000 \text{ kg}/m^3)(9.80 \text{ m}/s^2)(0.200 \text{ m}) = 1.05 \times 10^5 \text{ Pa} \quad \Diamond$

33. An empty rubber balloon has a mass of 0.012 0 kg. The balloon is filled with helium at a density of 0.181 kg / m³. At this density the balloon has a radius of 0.500 m. If the filled balloon is fastened to a vertical line, what is the tension in the line?

Solution

A free-body diagram of the balloon is shown in the sketch. The force **T** is the tension in the line, **B** is the buoyant force exerted on the balloon by the air surrounding it, and the total weight of the balloon and the helium filling it is

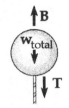

$$w_{total} = w_{balloon} + w_{helium}$$

Since the balloon is in equilibrium,　　$\Sigma F_y = +B - w_{total} - T = 0$

or the tension in the line is:　　$T = B - w_{total}$　　[1]

The weight of the balloon itself is

$$w_{balloon} = m_{balloon}g = (0.0120 \text{ kg})(9.80 \text{ m} / \text{s}^2) = 0.118 \text{ N}$$

The volume of helium is

$$V = \frac{4}{3}\pi R_{balloon}^3 = \frac{4}{3}\pi(0.500 \text{ m})^3 = 0.524 \text{ m}^3$$

and　　$w_{helium} = \rho_{helium} gV = (0.181 \text{ kg} / \text{m}^3)(9.80 \text{ m} / \text{s}^2)(0.524 \text{ m}^3) = 0.929 \text{ N}$

From Archimedes's principle, the buoyant force experienced by the balloon is equal to the weight of the air it displaces.

This is:　　$B = m_{air}g = (\rho_{air}V)g = (1.29 \text{ kg} / \text{m}^3)(0.524 \text{ m}^3)(9.80 \text{ m} / \text{s}^2) = 6.62 \text{ N}$

Equation [1] then gives the tension in the line holding the balloon as

$$T = 6.62 \text{ N} - (0.118 \text{ N} + 0.929 \text{ N}) = 5.57 \text{ N} \qquad \Diamond$$

39. A 1.00-kg beaker containing 2.00 kg of oil (density $= 916 \text{ kg} / \text{m}^3$) rests on a scale. A 2.00-kg block of iron is suspended from a spring scale and completely submerged in the oil (Fig. P9.39). Find the equilibrium readings of both scales.

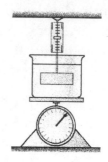

Figure P9.39

Solution   The volume of the iron block is

$$V = \frac{m_{\text{iron}}}{\rho_{\text{iron}}} = \frac{2.00 \text{ kg}}{7.86 \times 10^3 \text{ kg} / \text{m}^3} = 2.54 \times 10^{-4} \text{ m}^3$$

and the buoyant force exerted on the iron by the oil is

$$B = (\rho_{\text{oil}} V)g = (916 \text{ kg} / \text{m}^3)(2.54 \times 10^{-4} \text{ m}^3)(9.80 \text{ m} / \text{s}^2) = 2.28 \text{ N}$$

Applying $\Sigma F_y = 0$ to the iron block gives the support force exerted by the upper scale (and hence the reading on that scale) as

$$F_{\text{upper}} = m_{\text{iron}} g - B = 19.6 \text{ N} - 2.28 \text{ N} = 17.3 \text{ N} \qquad \lozenge$$

From Newton's third law, the iron exerts a reaction force of magnitude B downward on the oil (and hence the beaker). Other vertical forces acting on the beaker are (1) the combined weight of the beaker and oil, and (2) an upward support force exerted by the lower scale. Applying $\Sigma F_y = 0$ to the system consisting of the beaker and the oil gives

$$F_{\text{lower}} - B - (m_{\text{oil}} + m_{\text{beaker}})g = 0$$

The support force exerted by the lower scale (and the lower scale reading) is then

$$F_{\text{lower}} = B + (m_{\text{oil}} + m_{\text{beaker}})g = 2.28 \text{ N} + [(2.00 + 1.00) \text{ kg}](9.80 \text{ m} / \text{s}^2) = 31.7 \text{ N} \qquad \lozenge$$

42. A liquid ($\rho = 1.65 \text{ g} / \text{cm}^3$) flows through two horizontal sections of tubing joined end to end. In the first section the cross-sectional area is 10.0 cm², the flow speed is 275 cm/s, and the pressure is 1.20×10^5 Pa. In the second section the cross-sectional area is 2.50 cm². Calculate the smaller section's (a) flow speed and (b) pressure.

Solution

The situation described is illustrated in the sketch at the right. The given values are:

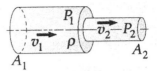

$$P_1 = 1.20 \times 10^5 \text{ Pa} \qquad v_1 = 275 \text{ cm / s} = 2.75 \text{ m / s}$$

$$A_1 = 10.0 \text{ cm}^2 \qquad A_2 = 2.50 \text{ cm}^2$$

The density of the flowing liquid is

$$\rho = \left(1.65 \text{ g / cm}^3\right)\left(\frac{1 \text{ kg}}{1\,000 \text{ g}}\right)\left(\frac{100 \text{ cm}}{1 \text{ m}}\right)^3 = 1.65 \times 10^3 \text{ kg / m}^3 \qquad \Diamond$$

(a) From the equation of continuity, $A_2 v_2 = A_1 v_1$, the flow speed in the second section of tubing is:

$$v_2 = \frac{A_1}{A_2} v_1 = \left(\frac{10.0 \text{ cm}^2}{2.50 \text{ cm}^2}\right)(2.75 \text{ m / s}) = 11.0 \text{ m / s} \qquad \Diamond$$

(b) Bernoulli's equation — a consequence of the principle of conservation of energy — states that $P + \frac{1}{2}\rho v^2 + \rho g y = \text{constant}$, where P is the pressure exerted by the fluid, ρ is the fluid density, v is the flow speed, and y is the elevation above some reference level. Consider two points on the center line of the tubes, one point in the larger tube and one point in the smaller tube. Then Bernoulli's equation requires that

$$P_2 + \frac{1}{2}\rho v_2^2 + \rho g y_2 = P_1 + \frac{1}{2}\rho v_1^2 + \rho g y_1$$

Since both sections of tubing are horizontal, $y_2 = y_1$ and the gravitational potential energy terms cancel. The pressure in the smaller tube is therefore

$$P_2 = P_1 + \frac{1}{2}\rho\left(v_1^2 - v_2^2\right)$$

or $\qquad P_2 = 1.20 \times 10^5 \text{ Pa} + \frac{1}{2}\left(1650 \text{ kg / m}^3\right)\left[(2.75 \text{ m / s})^2 - (11.0 \text{ m / s})^2\right]$

Thus, $P_2 = 2.64 \times 10^4 \text{ Pa}$ $\qquad \Diamond$

45. A jet of water squirts out horizontally from a hole near the bottom of the tank in Figure P9.45. If the hole has a diameter of 3.50 mm, what is the height, h, of the water level in the tank?

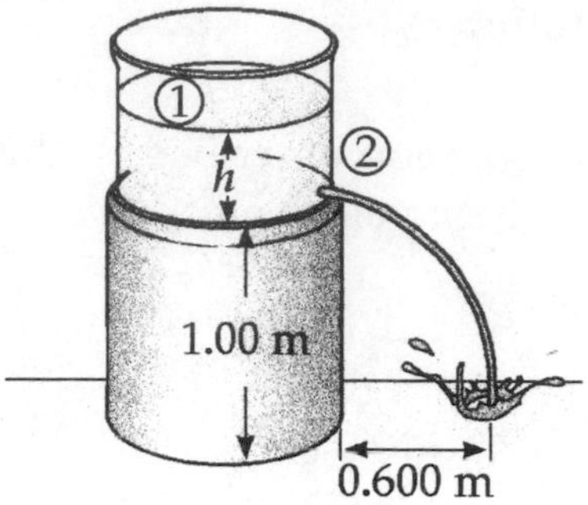

Figure P9.45

Solution

First, consider the projectile motion of the water droplets as they go from point 2 to the ground.

From $\Delta y = v_{iy}t + \frac{1}{2}a_y t^2$ with $v_{iy} = 0$

we find the time of flight as $t = \sqrt{\frac{2(\Delta y)}{a_y}} = \sqrt{\frac{2(-1.00 \text{ m})}{-9.80 \text{ m}/s^2}} = 0.452 \text{ s}$ ◊

The speed of the water as it emerges from the hole at point 2 may be determined from the horizontal motion:

$$v_2 = v_{ix} = \frac{\Delta x}{t} = \frac{0.600 \text{ m}}{0.452 \text{ s}} = 1.33 \text{ m}/s$$

We now use Bernoulli's equation, $P_1 + \frac{1}{2}\rho v_1^2 + \rho g y_1 = P_2 + \frac{1}{2}\rho v_2^2 + \rho g y_2$

with point 1 at the upper surface of the water in the tank and point 2 at hole. The tank is open to the atmosphere at both points 1 and 2,

so $P_1 = P_2 = P_{atm}$

We assume the tank is large enough that the speed of the water is negligible at point 1 in comparison to the speed of the emerging water (i.e., $v_1 \cong 0$).

Then, Bernoulli's equation reduces to $\rho g y_1 = \frac{1}{2}\rho v_2^2 + \rho g y_2$

and the height of the water level in the tank is $h = y_1 - y_2 = \frac{v_2^2}{2g} = \frac{(1.33 \text{ m}/s)^2}{2(9.80 \text{ m}/s^2)}$

$$h = 9.00 \times 10^{-2} \text{ m} = 9.00 \text{ cm}$$ ◊

Note that knowledge of the diameter of the hole is not needed for this solution.

54. Whole blood has a surface tension of $0.058 \, \text{N} / \text{m}$ and a density of $1\,050 \, \text{kg} / \text{m}^3$. To what height can whole blood rise in a capillary blood vessel that has a radius of 2.0×10^{-6} m if the contact angle is zero?

Solution The surface tension γ of a fluid is defined as the tension force per unit length (tangential to the fluid surface and tending to cause that surface to contract) along any line drawn on the surface of the fluid.

Consider the line along which the upper surface of a fluid in a capillary tube meets the wall of that tube as shown in the sketch. This line has a length equal to the circumference of the tube (i.e., $L = 2\pi r$). The surface (and hence the tension force) makes an angle ϕ, known as the contact angle, with the vertical wall of the tube at points on this line. The total upward force the tube wall exerts on the fluid is then

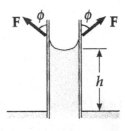

$$F = \gamma L \cos\phi = 2\pi \gamma \, r \cos\phi$$

The fluid then rises until the weight of the lifted fluid, $w = mg = \rho V g = \rho\left(\pi r^2 h\right)g$ equals the **upward** force F. When equilibrium is reached, we then have

$$\rho\left(\pi r^2 h\right)g = 2\pi \gamma \, r \cos\phi \qquad \text{or} \qquad h = \frac{2\gamma \cos\phi}{\rho g r}$$

If the contact angle is zero when whole blood rises in a capillary blood vessel, the height to which blood will rise in a vessel with a radius of $r = 2.0 \times 10^{-6}$ m is:

$$h = \frac{2(0.058 \, \text{N} / \text{m})\cos 0°}{\left(1\,050 \, \text{kg} / \text{m}^3\right)\left(9.80 \, \text{m} / \text{s}^2\right)\left(2.0 \times 10^{-6} \, \text{m}\right)} = 5.6 \, \text{m} \qquad\qquad \Diamond$$

63. Glucose, during transfusion, flows from a bag through a needle that is 3.0 cm long and has an inside diameter of 0.50 mm. If the transfusion rate is to be 0.50 L in 40 min, how high above the needle must the bag be held? Assume a density of $1\,040 \, \text{kg} / \text{m}^3$ and a coefficient of viscosity of $2.25 \times 10^{-3} \, \text{N} \cdot \text{s} / \text{m}^2$ for the glucose. Assume the patient's blood is nearly at atmospheric pressure.

Solution Assuming the pressure inside the vein is $P_2 \cong P_{atm}$, Poiseuille's law gives the required gauge pressure of the glucose as it enters the needle as

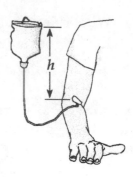

$$P_{gauge} = P_1 - P_{atm} = \frac{8\eta L(\text{flow rate})}{\pi R^4}$$

where η is the viscosity of the fluid flowing through the needle, L is the length of the needle, and R is its inner radius. If the patient is to be given 0.50 L of glucose in an elapsed time of 40 min, the required flow rate is:

$$\text{flow rate} = \left(\frac{0.50 \text{ L}}{40 \text{ min}}\right)\left(\frac{1 \text{ min}}{60 \text{ s}}\right)\left(\frac{1 \text{ m}^3}{10^3 \text{ L}}\right) = 2.1\times 10^{-7} \text{ m}^3/\text{s}$$

When the surface of the glucose (of density ρ) in the bag is height h above the entrance to the needle, the gauge pressure as the glucose enters the needle is $P_1 - P_{atm} = \rho g h$.

Poiseuille's law then becomes $\qquad\qquad \rho g h = \dfrac{8\eta L(\text{flow rate})}{\pi R^4}$

The required height of the glucose bag is then

$$h = \frac{8\eta L(\text{flow rate})}{\pi \rho g R^4} = \frac{8(2.25\times 10^{-3} \text{ N}\cdot\text{s}/\text{m}^2)(3.0\times 10^{-2} \text{ m})(2.1\times 10^{-7} \text{ m}^3/\text{s})}{\pi(1040 \text{ kg}/\text{m}^3)(9.80 \text{ m}/\text{s}^2)(0.25\times 10^{-3} \text{ m})^4}$$

or $\qquad h = 0.90 \text{ m} = 90 \text{ cm}$ $\qquad\qquad\qquad\qquad\qquad\qquad\qquad\qquad\qquad\qquad\qquad\qquad$ ◊

---

70. Small spheres of diameter 1.00 mm fall through 20 °C water with a terminal speed of 1.10 cm/s. Calculate the density of the spheres.

Solution As a sphere falls through the water, three forces act on it:

(1) The weight of the sphere, $w = mg = \rho_s V g = \rho_s g\left(\frac{4}{3}\pi r^3\right)$ where ρ_s is the density of the sphere;

(2) A buoyant force, equal to the weight of the displaced water, $B = m_{water} g = \rho_w V g = \rho_w g\left(\frac{4}{3}\pi r^3\right)$ where ρ_w is the water's density;

(3) A viscous resistance force given by Stoke's law as $F = 6\pi\eta r v$. Here η is the viscosity of the water and v is the speed of the falling sphere.

The sphere reaches terminal speed (i.e., ceases to accelerate) when the net force acting on it becomes zero. Thus, at the terminal speed $(v = v_t)$, the total upward force, F and B, and the downward gravitational force, w, are equal.

That is, $F + B = w$: $\qquad 6\pi\eta r v_t + \rho_w g\left(\frac{4}{3}\pi r^3\right) = \rho_s g\left(\frac{4}{3}\pi r^3\right)$

which reduces to $\qquad \rho_s = \rho_w + \dfrac{9\eta v_t}{2r^2 g}$ $\hfill$ [1]

Given values of $\qquad \eta = 1.00 \times 10^{-3}$ N·s / m^2 $\hfill$ (Table 9.5)

$\qquad\qquad\qquad\quad \rho_w = 1\,000$ kg / m^3 $\hfill$ (Table 9.3)

then $\qquad r = \frac{1}{2}(\text{diameter}) = \frac{1}{2}\left(1.00 \times 10^{-3} \text{ m}\right) = 5.00 \times 10^{-4}$ m

and $\qquad v_t = 1.10$ cm / s $= 1.10 \times 10^{-2}$ m / s

Equation [1] becomes $\qquad \rho_s = 1000 \text{ kg / m}^3 + \dfrac{9\left(1.00 \times 10^{-3} \text{N·s / m}^2\right)\left(1.10 \times 10^{-2} \text{m / s}\right)}{2\left(5.00 \times 10^{-4} \text{ m}\right)^2\left(9.80 \text{ m / s}^2\right)}$

Thus, the density of the sphere is $\quad \rho_s = 1.02 \times 10^3$ kg / m^3 $\hfill \Diamond$

---

73. The approximate inside diameter of the aorta is 0.50 cm; that of a capillary is 10 μm. The approximate average blood flow speed is 1.0 m/s in the aorta and 1.0 cm/s in the capillaries. If all the blood in the aorta eventually flows through the capillaries, estimate the number of capillaries in the circulatory system.

Solution If the diameters of the aorta and of a capillary are d_1 and d_2, respectively, their cross-sectional areas are

$$A_{\text{aorta}} = A_1 = \frac{\pi d_1^{\,2}}{4} \qquad \text{and} \qquad A_{\text{capillary}} = A_c = \frac{\pi d_2^{\,2}}{4}$$

Assuming the circulatory system has a total of N capillaries, the total cross-sectional area carrying blood from the aorta is $A_2 = N A_c = N(\pi d_2^{\,2}) / 4$. The equation of continuity then requires that $A_2 v_2 = A_1 v_1$, where v_1 is the blood flow speed in the aorta and v_2 is the flow speed in a capillary.

This gives $\qquad N\left(\dfrac{\pi d_2^{\,2}}{4}\right)v_2 = \left(\dfrac{\pi d_2^{\,2}}{4}\right)v_1$

so the number of capillaries in the circulatory system must be

$$N = \left(\frac{d_1}{d_2}\right)^2 \left(\frac{v_1}{v_2}\right) = \left(\frac{0.50 \times 10^{-2} \text{ m}}{10 \times 10^{-6} \text{ m}}\right)^2 \left(\frac{1.0 \text{ m/s}}{1.0 \times 10^{-2} \text{ m/s}}\right) = 2.5 \times 10^7 = 25 \text{ million } \Diamond$$

88. Oil having a density of 930 kg/m^3 floats on water. A rectangular block of wood 4.00 cm high and with a density of 960 kg/m^3 floats partly in the oil and partly in the water. The oil completely covers the block. How far below the interface between the two liquids is the bottom of the block?

Solution   Assume the block floats in equilibrium with the bottom of the block a distance x below the oil-water interface. The top of the block is a distance $(4.00 \text{ cm} - x)$ above the interface as shown in the sketch. At equilibrium, the total buoyant force equals the weight of the block:

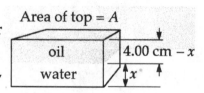

$$w = B_{\text{total}} = B_{\text{oil}} + B_{\text{water}} \tag{1}$$

The block's weight is $w = mg = \rho_{\text{wood}}(\text{Volume of block})g = \rho_{\text{wood}} A(4.00 \text{ cm})g$

and the individual buoyant forces are

B_{oil} = weight of displaced oil = $\rho_{\text{oil}}(\text{volume of displaced oil})g$

and B_{water} = weight of displaced water = $\rho_{\text{water}}(\text{volume of displaced water})g$

Substituting these values into our buoyancy equation, Equation [1],

$$\rho_{\text{wood}} A(4.00 \text{ cm})g = \rho_{\text{oil}}[A(4.00 \text{ cm} - x)]g + \rho_{\text{water}}[Ax]g$$

Cancelling the cross-sectional area A and the acceleration of gravity g,

$$(\rho_{\text{water}} - \rho_{\text{oil}})x = (\rho_{\text{wood}} - \rho_{\text{oil}})(4.00 \text{ cm})$$

Since $\rho_{\text{water}} = 1000 \text{ kg/m}^3$, $\rho_{\text{oil}} = 930 \text{ kg/m}^3$,   and $\rho_{\text{wood}} = 960 \text{ kg/m}^3$

we can solve for the x, the distance the block's bottom is below the oil-water interface:

$$x = \left[\frac{\rho_{\text{wood}} - \rho_{\text{oil}}}{\rho_{\text{water}} - \rho_{\text{oil}}}\right](4.00 \text{ cm}) = \left[\frac{30 \text{ kg/m}^3}{70 \text{ kg/m}^3}\right](4.00 \text{ cm}) = 1.7 \text{ cm} \qquad \Diamond$$

Chapter 10
THERMAL PHYSICS

NOTES ON SELECTED CHAPTER SECTIONS

10.1 Temperature and the Zeroth Law of Thermodynamics

The zeroth law of thermodynamics (or the equilibrium law) can be stated as follows:

> If bodies A and B are separately in thermal equilibrium with a third body, C, then A and B will be in thermal equilibrium with each other if placed in thermal contact.

Two objects in thermal equilibrium with each other are at the same temperature.

10.2 Thermometers and Temperature Scales

The physical property used in a constant volume gas thermometer is the pressure variation with temperature of a fixed volume of gas. The temperature readings are nearly independent of the substance used in the thermometer.

The **triple point of water**, which is **the single temperature and pressure at which water, water vapor, and ice can coexist in equilibrium,** was chosen as a convenient and reproducible reference temperature for the Kelvin scale. It occurs at a temperature of 0.01 °C and a pressure of 4.58 mm of mercury. The temperature at the triple point of water on the Kelvin scale has been assigned a value of 273.16 kelvins (K). Thus, the SI unit of temperature, the kelvin, is defined as **1/273.16 of the temperature of the triple point of water**.

The temperature 0 K is often referred to as absolute zero although this temperature has never been achieved.

10.3 Thermal Expansion of Solids and Liquids

Liquids generally increase in volume with increasing temperature and have volume expansion coefficients about ten times greater than those of solids. Water is an exception to this rule; as the temperature increases from 0 °C to 4 °C, water contracts and thus its density increases. Above 4 °C, water expands with increasing temperature. In other words, the density of water reaches its maximum value (1000 kg/m³) 4 degrees above the freezing point.

10.5 Avogadro's Number and the Ideal Gas Law

Equal volumes of gas at the same temperature and pressure contain the same numbers of molecules.

One-mole quantities of all gases at standard temperature and pressure contain the same numbers of molecules.

10.6 The Kinetic Theory of Gases

A microscopic **model of an ideal gas** is based on the following assumptions:

1. **The number of molecules is large, and the average separation between them is large** compared with their dimensions. Therefore, the molecules occupy a negligible volume compared with the volume of the container.

2. **The molecules obey Newton's laws of motion, but the individual molecules move in a random fashion.** By random fashion, we mean that the molecules move in all directions with equal probability and with various speeds. This distribution of velocities does not change in time, despite the collisions between molecules.

3. **The molecules undergo elastic collisions with each other.** Thus, the molecules are considered to be structureless (that is, point masses), and in the collisions both kinetic energy and momentum are conserved.

4. **The forces between molecules are negligible, except during a collision.** The forces between molecules are short-range, so that the only time the molecules interact with each other is during a collision.

5. **The gas under consideration is a pure gas.** That is, all molecules are identical.

6. **The molecules of the gas make perfectly elastic collisions with the walls of the container.** Hence, the wall will eject as many molecules as it absorbs, and the ejected molecules will have the same average kinetic energy as the absorbed molecules.

EQUATIONS AND CONCEPTS

T_C is the Celsius temperature and T is the Kelvin temperature (sometimes called the absolute temperature). The size of a degree on the Kelvin scale is identical to the size of a degree on the Celsius scale.

$$T_C = T - 273.15 \qquad (10.1)$$

These equations are useful in converting temperature values between Fahrenheit and Celsius scales.

$$T_F = \frac{9}{5} T_C + 32 \qquad (10.2)$$

$$T_C = \frac{5}{9}(T_F - 32)$$

The SI unit of temperature is the kelvin, K, which is defined as $1/273.16$ of the temperature of the triple point of water. The notations °C and °F refer to actual temperature values in degrees Celsius and degrees Fahrenheit.

Comment on units and notation

These are two forms of the basic equation for the thermal expansion of a solid. The change in length is proportional to the original length and to the change in temperature. The constant α is characteristic of a particular type of material and is called the average temperature coefficient of linear expansion.

$$\Delta L = \alpha L_0 \Delta T \qquad (10.4)$$

$$L = L_0(1 + \alpha \Delta T)$$

If a body's temperature changes, its surface area and its volume will change by amounts which are proportional to the changes in temperature. γ (gamma) is the average temperature coefficient of area expansion; β (beta) is the average temperature coefficient of volume expansion.

$$\Delta A = \gamma A_0 \Delta T \qquad (10.5)$$

$$\Delta V = \beta V_0 \Delta T \qquad (10.6)$$

$$\gamma \cong 2\alpha$$

$$\beta \cong 3\alpha$$

The number of moles in a sample of an element or compound is the ratio of the mass, m, of the sample to the atomic or molar mass of the material. **Also, one mole of a substance contains Avogadro's number of molecules.**

$$n = \frac{m}{\text{molar mass}} \qquad (10.7)$$

$$N_A = 6.02 \times 10^{23} \frac{\text{particles}}{\text{mol}}$$

$$m_{\text{atom}} = \frac{\text{molar mass}}{N_A}$$

Boyle's law states that when a gas is maintained at constant temperature, its pressure is inversely proportional to its volume.

$$P \propto \left(\frac{1}{V}\right); \quad T = \text{constant}$$

Charles's Law states that when constant pressure is maintained, a gas's volume is directly proportional to its absolute temperature.

$$V \propto T; \quad P = \text{constant}$$

This is the equation of state of an ideal gas. In this equation, T must be the **temperature in kelvins.** In this equation n **is the number of moles of gas in the sample** and R is the universal gas constant. R must be expressed in units which are consistent with those used for pressure P and volume V.

$$PV = nRT \qquad (10.8)$$

$$R = 8.31 \text{ J / mol} \cdot \text{K} \qquad (10.9)$$

$$R = 0.0821 \text{ L} \cdot \text{atm / mol} \cdot \text{K}$$

The number of moles in a sample of gas equals the number of molecules, N, divided by Avogadro's number, N_A.

$$n = \frac{N}{N_A} \qquad (10.10)$$

The ideal gas law can also be expressed in this alternate form where N **is the total number of molecules** in the sample and k_B is Boltzmann's constant.

$$PV = Nk_BT \qquad (10.11)$$

$$k_B = \frac{R}{N_A} = 1.38 \times 10^{-23} \text{ J / K} \qquad (10.12)$$

In an ideal gas, the pressure of the gas is proportional to the number of molecules per unit volume and proportional to the average kinetic energy of the molecules.

$$P = \frac{2}{3}\left(\frac{N}{V}\right)\left(\frac{1}{2}m\overline{v^2}\right) \qquad (10.13)$$

Equations 10.11 and 10.13 can be combined to present an important result: the average translational kinetic energy per molecule is directly proportional to the absolute temperature of the gas.

$$\frac{1}{2}m\overline{v^2} = \frac{3}{2}k_B T \qquad (10.15)$$

This expression for the root mean square (rms) speed shows that, at a given temperature, lighter molecules move faster on the average than heavier ones. In this equation m is the molecular mass and M is the molar mass.

$$v_{rms} = \sqrt{\overline{v^2}} = \sqrt{\frac{3k_B T}{m}} = \sqrt{\frac{3RT}{M}} \qquad (10.18)$$

REVIEW CHECKLIST

▷ Describe the operation of the constant-volume gas thermometer and how it is used to determine the Kelvin temperature scale. Convert between the various temperature scales, especially the conversion from degrees Celsius into kelvins, degrees Fahrenheit into kelvins, and degrees Celsius into degrees Fahrenheit.

▷ Define the linear expansion coefficient and volume expansion coefficient for an isotropic solid, and understand how to use these coefficients in practical situations involving expansion or contraction.

▷ Understand the assumptions made in developing the molecular model of an ideal gas; and apply the equation of state for an ideal gas to calculate pressure, volume, temperature, or number of moles.

▷ Define each of the following terms: **molecular weight, mole, Avogadro's number, universal gas constant, and Boltzmann's constant.**

SOLUTIONS TO SELECTED END-OF-CHAPTER PROBLEMS

3. Convert the following temperatures to their values on the Fahrenheit and Kelvin scales: (a) the boiling point of liquid hydrogen −252.87 °C; (b) the temperature of a room at 20 °C.

Solution

The Celsius and Kelvin temperature scales have the same size scale divisions, with the only difference being that the zero point on the Kelvin scale is 273.15 scale divisions below the zero point on the Celsius scale. Hence, the conversion from Celsius temperature readings to the corresponding Kelvin temperature is

$$T_K = T_C + 273.15 \qquad\qquad [1]$$

The Celsius scale divisions are 9/5 (or almost double) the size of the scale divisions on the Fahrenheit thermometer.

Also, the zero point on the Celsius thermometer is 32 Fahrenheit divisions above the zero point on the Fahrenheit thermometer (water freezes at 0 °C and 32 °F). Thus, the conversion from Celsius to Fahrenheit is

$$T_F = \frac{9}{5}T_C + 32 \;°F \qquad\qquad [2]$$

(a) Using Equations [1] and [2], the Fahrenheit and Kelvin temperatures corresponding to a Celsius temperature of $T_C = -252.87$ °C are

$$T_F = \frac{9}{5}(-252.87) + 32.0 \;°F = -423 \;°F$$

and $\quad T_K = -252.87 + 273.15 = 20.28$ K $\qquad\qquad$ ◊

(b) For a Celsius temperature of $T_C = 20$ °C, the conversions give

$$T_F = \frac{9}{5}(20) + 32.0 \;°F = 68 \;°F$$

and $\quad T_K = 20 + 273.15 = 293$ K $\qquad\qquad$ ◊

6. A constant-volume gas thermometer is calibrated in dry ice (–80.0 °C) and in boiling ethyl alcohol (78.0 °C). The two pressures are 0.900 atm and 1.635 atm. (a) What value of absolute zero does the calibration yield? (b) What pressures would be found at the freezing and boiling points of water? (Note that we have a linear relationship between P and T as $P = A + BT$, where A and B are constants.)

Solution When the volume occupied by a low density gas is constant, the graph of pressure versus temperature is a straight line. This linear relation is summarized by an equation of the form $P = A + BT$, where A and B are constants. To determine the values of these constants for this gas thermometer, use the two calibration points. The first calibration point ($P = 0.900$ atm at $T = -80.0$ °C) yields the equation:

$$0.900 \text{ atm} = A - (80.0 \text{ °C})B \qquad \text{[1]}$$

The second calibration point ($P = 1.635$ atm at $T = 78.0$ °C) gives the equation:

$$1.635 \text{ atm} = A + (78.0 \text{ °C})B \qquad \text{[2]}$$

Subtracting Equation [1] from Equation [2] gives

$$(158.0 \text{ °C})B = 0.735 \text{ atm} \qquad \text{or} \qquad B = 4.65 \times 10^{-3} \text{ atm/ °C}$$

Substituting this result into either Equation [1] or Equation [2] then yields $A = 1.27$ atm. Thus, the equation of the calibration curve for this constant-volume gas thermometer is

$$P = 1.27 \text{ atm} + \left(4.65 \times 10^{-3} \text{ atm/ °C}\right)T$$

(a) Gas at absolute zero would exert zero pressure. The calibration curve at $P = 0$ yields:

$$T = -1.27 \text{ atm}\left(\frac{1}{4.65 \times 10^{-3} \text{ atm/ °C}}\right) = -273 \text{ °C} \qquad \Diamond$$

(b) At the freezing point of water, $T = 0$ °C and the calibration curve predicts a pressure of

$$P_f = 1.27 \text{ atm} \qquad \Diamond$$

At the boiling point of water ($T = 100$ °C), the calibration curve yields:

$$P_b = 1.27 \text{ atm} + \left(4.65 \times 10^{-3} \text{ atm/ °C}\right)(100 \text{ °C}) = 1.74 \text{ atm} \qquad \Diamond$$

15. A brass ring of diameter 10.00 cm at 20.0 °C is heated and slipped over an aluminum rod of diameter 10.01 cm at 20.0 °C. Assuming the average coefficients of linear expansion are constant, (a) to what temperature must this combination be cooled to separate them? Is this attainable? (b) What if the aluminum rod were 10.02 cm in diameter?

Solution The magnitude, L, of a linear dimension on an object at temperature T is given by

$$L = L_0[1 + \alpha(T - T_0)] = L_0 + \alpha L_0(T - T_0)$$

where L_0 is the magnitude of that dimension at temperature T_0 and α is the coefficient of linear expansion of the material making up the object. To remove the ring from the rod, the diameter of the ring must be at least as large as the diameter of the rod.

Thus, we require $\left(L_f\right)_{\text{brass}} = \left(L_f\right)_{\text{Al}}$

or $$L_{\text{brass}} + \alpha_{\text{brass}} L_{\text{brass}}(\Delta T) = L_{\text{Al}} + \alpha_{\text{Al}} L_{\text{Al}}(\Delta T)$$

This gives $$\Delta T = \frac{L_{\text{Al}} - L_{\text{brass}}}{\alpha_{\text{brass}} L_{\text{brass}} - \alpha_{\text{Al}} L_{\text{Al}}}$$

(a) If $L_{\text{Al}} = 10.01$ cm and $L_{\text{brass}} = 10.00$ cm at $T = 20.0$ °C, the change in temperature required to make the new lengths equal is

$$\Delta T = \frac{10.01 - 10.00}{\left[19 \times 10^{-6} \ (°C)^{-1}\right](10.00) - \left[24 \times 10^{-6} \ (°C)^{-1}\right](10.01)} = -199 \ °C$$

The final temperature will then be

$$T_f = T_0 + \Delta T = 20.0 \ °C - 199 \ °C = -179 \ °C \quad \text{which is attainable.} \qquad \lozenge$$

(b) If $L_{\text{Al}} = 10.02$ cm and $L_{\text{brass}} = 10.00$ cm at $T = 20.0$ °C, the change in temperature required to make the new lengths equal is

$$\Delta T = \frac{10.02 - 10.00}{\left[19 \times 10^{-6} \ (°C)^{-1}\right](10.00) - \left[24 \times 10^{-6} \ (°C)^{-1}\right](10.02)} = -396 \ °C$$

The final temperature will then be

$$T_f = T_0 + \Delta T = 20.0 \ °C - 396 \ °C = -376 \ °C$$

which is below absolute zero (−273.15 °C) and, therefore, unattainable. $\qquad \lozenge$

21. An automobile fuel tank is filled to the brim with 45 L (12 gal) of gasoline at 10 °C. Immediately afterward, the vehicle is parked in the sun, where the temperature is 35 °C. How much gasoline overflows from the tank as a result of the expansion? (Neglect the expansion of the tank.)

Solution If a solid or liquid occupies volume V_0 at temperature T_0, its volume at temperature T will be given by

$$V = V_0\big[1 + \beta(T - T_0)\big] = V_0 + \beta V_0(T - T_0)$$

where β is the coefficient of volume expansion of that material. Thus, the change in volume that occurs is

$$\Delta V = V - V_0 = \beta V_0(T - T_0)$$

The coefficient of volume expansion for gasoline is $\beta = 9.6 \times 10^{-4}$ $(°C)^{-1}$. Therefore, if the 45-L fuel tank is filled to the brim with gasoline at $T_0 = 10$ °C, the increase in volume (and hence the amount of overflow) as the temperature rises to $T = 35$ °C will be

$$\Delta V = \big[9.6 \times 10^{-4} \ (°C)^{-1}\big](45 \ \text{L})(35 \ °C - 10 \ °C) = 1.1 \ \text{L} \qquad \Diamond$$

29. (a) An ideal gas occupies a volume of 1.0 cm^3 at 20 °C and atmospheric pressure. Determine the number of molecules of gas in the container. (b) If the pressure of the $1.0 \text{-} cm^3$ volume is reduced to 1.0×10^{-11} Pa (an extremely good vacuum) while the temperature remains constant, how many moles of gas remain in the container?

Solution

(a) The ideal gas law may be written as $PV = nRT$ where P is the absolute pressure (in Pa) of the gas, V is the volume (in m^3) the gas occupies, T is the Kelvin temperature, n is the number of moles of gas present, and $R = 8.31$ J/mol · K is the universal gas constant.

For the given gas, $V = 1.0 \ cm^3\big[1 \ \text{m}/100 \ \text{cm}\big]^3 = 1.0 \times 10^{-6} \ m^3$

$$T = 20 \ °C = 293 \ \text{K}$$

and $\qquad P = 1 \ \text{atm} = 1.013 \times 10^5 \ \text{Pa}$

Therefore, $\qquad n = \dfrac{PV}{RT} = \dfrac{\big(1.013 \times 10^5 \ \text{Pa}\big)\big(1.0 \times 10^{-6} \ m^3\big)}{(8.31 \ \text{J/mol} \cdot \text{K})(293 \ \text{K})} = 4.2 \times 10^{-5} \ \text{mol}$

This is the number of moles present; since Avogadro's number, $N_A = 6.02 \times 10^{23}$ mol^{-1}, gives the number of molecules in one mole of any gas, the number of molecules in the container is:

$$N = nN_A = \left(4.2 \times 10^{-5} \text{ mol}\right)\left(6.02 \times 10^{23} \text{ mol}^{-1}\right) = 2.5 \times 10^{19} \qquad \Diamond$$

(b)   The equation $P_1V_1 = n_1RT_1$ refers to the initial state of the gas,

and the equation $P_2V_2 = n_2RT_2$ describes the final state of the gas.

Dividing the second equation by the first, recognizing that the volume and temperature are both held constant in this case, gives $P_2/P_1 = n_2/n_1$.

Thus, the number of moles of gas remaining in the container is

$$n_2 = \left(\frac{P_2}{P_1}\right)n_1 = \left(\frac{1.0 \times 10^{-11} \text{ Pa}}{1.013 \times 10^5 \text{ Pa}}\right)\left(4.2 \times 10^{-5} \text{ mol}\right) = 4.1 \times 10^{-21} \text{ mol} \Diamond$$

34.   A cylindrical diving bell, 3.00 m in diameter and 4.00 m tall with an open bottom, is submerged to a depth of 220 m in the ocean. The surface temperature is 25.0 °C, and the temperature 220 m down is 5.00 °C. The density of sea water is 1025 kg / m^3. How high does the sea water rise in the bell when it is submerged?

Solution

At the surface of the ocean, air fills the entire volume of the diving bell. This initial volume is

$$V_1 = \pi r^2 h_1$$

where $h_1 = 4.00$ m is the height of the cylindrical space inside the bell. As the bell is lowered into the ocean, the trapped air is subject to the prevailing water pressure since the diving bell is open at the bottom. The pressure at a depth $H = 220$ m in the ocean

is $P_2 = P_{atm} + \rho_{sea\ water}gH$

or $P_2 = 1.013 \times 10^5 \text{ Pa} + \left(1025 \text{ kg /m}^3\right)\left(9.80 \text{ m/s}^2\right)(220 \text{ m}) = 2.31 \times 10^6 \text{ Pa}$

The initial state of the air is described by $\qquad P_1V_1 = n_1RT_1$

and the final state by $\qquad P_2V_2 = n_2RT_2$

Since no air enters or leaves the bell as it is lowered, the number of moles of air present is constant $(n_2 = n_1)$.

Dividing the two state equations then gives $\qquad \dfrac{P_2V_2}{P_1V_1} = \dfrac{T_2}{T_1}$

or $\qquad V_2 = \left(\dfrac{P_1}{P_2}\right)\left(\dfrac{T_2}{T_1}\right)V_1$

Recognizing that the radius of the diving bell remains constant, the new height of the cylindrical space filled by the trapped air is

$$h_2 = \left(\dfrac{1}{\pi r^2}\right)V_2 = \left(\dfrac{1}{\pi r^2}\right)\left(\dfrac{P_1}{P_2}\right)\left(\dfrac{T_2}{T_1}\right)\pi r^2 h_1 = \left(\dfrac{P_1}{P_2}\right)\left(\dfrac{T_2}{T_1}\right)h_1$$

Thus, with initial and final temperatures of $\qquad T_1 = 25.0\ °C + 273 = 298\ K$

and $\qquad T_2 = 5.00\ °C + 273 = 278\ K$

the final height of the trapped air space is

$$h_2 = \left(\dfrac{1.013\times10^5\,\text{Pa}}{2.31\times10^6\ \text{Pa}}\right)\left(\dfrac{278\ K}{298\ K}\right)(4.00\ m) = 0.164\ m$$

Therefore, as the bell is submerged, the sea water rises a distance of

$$d = h_1 - h_2 = 4.00\ m - 0.164\ m \doteq 3.84\ m \qquad \Diamond$$

---

41. A cylinder contains a mixture of helium and argon gas in equilibrium at a temperature of 150 °C. (a) What is the average kinetic energy of each type of molecule? (b) What is the rms speed of each type of molecule?

Solution

(a) The absolute temperatures of both the helium and argon gases are

$$T_K = T_C + 273 = 150 + 273 = 423\ K$$

Therefore, the average kinetic energy of a molecule in either gas is given by

$$\langle KE \rangle_{\text{molecule}} = \tfrac{3}{2}k_B T_K = \tfrac{3}{2}\left(1.38\times10^{-23}\ \text{J / K}\right)(423\ K) = 8.76\times10^{-21}\ \text{J} \qquad \Diamond$$

(b)   Since $\langle KE \rangle_{molecule} = \frac{1}{2}mv_{rms}^2$, where m is the mass of the molecule and v_{rms} is the root-mean-square speed of the molecules in that gas,

we have $\qquad v_{rms} = \sqrt{\dfrac{2\langle KE \rangle_{molecule}}{m}}$

For helium, $\qquad m = \dfrac{M}{N_A} = \dfrac{4.00 \times 10^{-3} \text{ kg / mol}}{6.02 \times 10^{23} \text{ molecules / mol}} = 6.64 \times 10^{-27} \text{ kg}$

and $\qquad v_{rms} = \sqrt{\dfrac{2(8.76 \times 10^{-21} \text{ J})}{6.64 \times 10^{-27} \text{ kg}}} = 1.62 \times 10^3 \text{ m / s} = 1.62 \text{ km / s}$ $\qquad\qquad \Diamond$

For argon, $\qquad m = \dfrac{M}{N_A} = \dfrac{39.9 \times 10^{-3} \text{ kg / mol}}{6.02 \times 10^{23} \text{ molecules / mol}} = 6.63 \times 10^{-26} \text{ kg}$

and $\qquad v_{rms} = \sqrt{\dfrac{2(8.76 \times 10^{-21} \text{ J})}{6.63 \times 10^{-26} \text{ kg}}} = 514 \text{ m / s}$ $\qquad\qquad\qquad\qquad \Diamond$

---

43.   Superman leaps in front of Lois Lane to save her from a volley of bullets. In a 1-min interval, an automatic weapon fires 150 bullets, each of mass 8.0 g, at 400 m/s. The bullets strike his mighty chest, which has an area of 0.75 m^2. Find the average force exerted on Superman's chest if the bullets bounce back after an elastic, head-on collision.

Solution   Choosing the positive direction to be directed from the shooter toward Superman, the initial velocity of a bullet is $\mathbf{v}_i = +400 \text{ m / s}$. Since the bullets undergo an elastic head-on collision, their final velocity is $\mathbf{v}_f = -400 \text{ m / s}$. The impulse experienced by each bullet is

$$\text{Impulse})_{bullet} = m\mathbf{v}_f - m\mathbf{v}_i$$

$$= (8.0 \times 10^{-3} \text{ kg})[-400 \text{ m / s} - (+400 \text{ m / s})] = -6.4 \text{ kg} \cdot \text{m / s}$$

From Newton's third law, each bullet then imparts an impulse of $+6.4 \text{ kg} \cdot \text{m / s}$ to Superman's chest. The total impulse he experiences in a time interval of $\Delta t = 1.0 \text{ min} = 60 \text{ s}$ is given by

$$\text{Impulse} = 150(+6.4 \text{ kg} \cdot \text{m / s}) = +9.6 \times 10^2 \text{ kg} \cdot \text{m / s}$$

and the average force exerted on his chest is

$$\overline{F} = \frac{\text{Impulse}}{\Delta t} = \frac{+9.6 \times 10^2 \text{ kg} \cdot \text{m} / \text{s}}{60 \text{ s}} = +16 \text{ N}$$

or $\overline{F} = 16 \text{ N}$ directed from the shooter toward Superman.   ◊

50. A vertical cylinder of cross-sectional area 0.050 m^2 is fitted with a tight-fitting, frictionless piston of mass 5.0 kg (Fig. P10.50). If there are 3.0 mol of an ideal gas in the cylinder at 500 K, determine the height h at which the piston will be in equilibrium under its own weight.

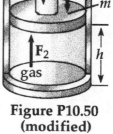

**Figure P10.50
(modified)**

Solution   Three forces act on the piston.

These are: (1) its own weight, $w = mg$, directed downward;

(2) an upward force $F_2 = P_g A$, where P_g is the gas pressure and A is the cross-sectional area of the piston;

and   (3) a downward force $F_3 = P_{atm} A$ exerted on the piston by the air above it.

When the piston is in equilibrium, $\sum F_y = F_2 - w - F_3 = 0$

so $P_g A = mg + P_{atm} A$ **[1]**

The pressure of the gas is given by the ideal gas law as $P_g = \dfrac{nRT}{V} = \dfrac{nRT}{Ah}$

where h is the height of the cylindrical volume of gas.   Substituting this into Equation [1] gives

$$\left(\frac{nRT}{Ah}\right) A = mg + P_{atm} A \qquad \text{or} \qquad \frac{1}{h} = \frac{mg + P_{atm} A}{nRT}$$

Thus, at equilibrium, the height of the trapped column of gas is:

$$h = \frac{nRT}{mg + P_{atm} A} = \frac{(3.0 \text{ mol})(8.31 \text{ J/mol} \cdot \text{K})(500 \text{ K})}{(5.0 \text{ kg})(9.8 \text{ m/s}^2) + (1.013 \times 10^5 \text{ Pa})(0.050 \text{ m}^2)} = 2.4 \text{ m} \quad ◊$$

51. A liquid with coefficient of volume expansion β just fills a spherical flask of volume V_0 at temperature T (Fig. P10.51). The flask is made of a material that has a coefficient of linear expansion α. The liquid is free to expand into a capillary of cross-sectional area A at the top. (a) If the temperature increases by ΔT, show that the liquid rises in the capillary by the amount $\Delta h = (V_0/A)(\beta - 3\alpha)\Delta T$. (b) For a typical system, such as a mercury thermometer, why is it a good approximation to neglect the expansion of the flask?

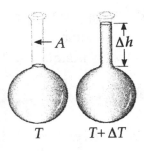

Figure P10.51

Solution

(a) When the temperature increases by ΔT, the volume of the liquid expands by an amount

$$\Delta V_{\text{liquid}} = \beta V_0(\Delta T)$$

and the volume enclosed by the spherical flask increases by the amount

$$\Delta V_{\text{flask}} = (3\alpha)V_0(\Delta T)$$

The volume of liquid which must overflow into the capillary is

$$V_{\text{overflow}} = \Delta V_{\text{liquid}} - \Delta V_{\text{flask}} = A(\Delta h)$$

where A is the cross-sectional area of the capillary and Δh is the height to which the liquid will rise in the capillary.

Therefore, $\qquad \Delta h = \dfrac{V_{\text{overflow}}}{A} = \dfrac{\beta V_0(\Delta T) - (3\alpha)V_0(\Delta T)}{A}$

$$\Delta h = \left(\frac{V_0}{A}\right)(\beta - 3\alpha)\Delta T \qquad\qquad \Diamond$$

(b) For a mercury thermometer, $\qquad \beta_{\text{Hg}} = 1.82 \times 10^{-4}\ (^\circ\text{C})^{-1}$

and (assuming Pyrex glass), $\qquad 3\alpha_{\text{glass}} = 3\left(3.2 \times 10^{-6}\ (^\circ\text{C})^{-1}\right) = 9.6 \times 10^{-6}\ (^\circ\text{C})^{-1}$

Thus, the expansion of the mercury is almost 20 times that of the glass flask. This means that it is a good approximation to neglect the expansion of the flask. $\qquad \Diamond$

56. A copper rod and steel rod are heated. At 0 °C the copper rod has a length of L_C the steel one has a length L_S. When the rods are being heated or cooled, a difference of 5.00 cm is maintained between their lengths. Determine the values of L_C and L_S.

Solution

If the difference in the lengths of the two rods is to remain constant as the rods expand or contract, any change in the length of one rod must be matched by an equal change in the length of the other rod. That is, it is necessary that $\Delta L_C = \Delta L_S$ for any change in temperature ΔT.

When a change in temperature ΔT occurs, the change in the lengths of the rods are

$$\Delta L_C = \alpha_C L_C(\Delta T) \qquad \text{and} \qquad \Delta L_S = \alpha_S L_S(\Delta T)$$

where α_C and α_S are the coefficients of linear expansion for copper and steel, respectively.

Thus, we must require that $\qquad\qquad\qquad \alpha_C L_C(\Delta T) = \alpha_S L_S(\Delta T)$

or $\qquad\qquad\qquad L_C = (\alpha_S / \alpha_C) L_S$ **[1]**

We must also impose the requirement that $\qquad L_S - L_C = 5.00 \text{ cm}$ **[2]**

Substituting Equation [1] into Equation [2] yields $\quad L_S\left(1 - \dfrac{\alpha_S}{\alpha_C}\right) = 5.00 \text{ cm}$

or $\qquad\qquad\qquad L_S = \dfrac{5.00 \text{ cm}}{1 - \alpha_S / \alpha_C}$

Since $\qquad\qquad\qquad \alpha_C = 17 \times 10^{-6} \ (°\text{C})^{-1}$

and $\qquad\qquad\qquad \alpha_S = 11 \times 10^{-6} \ (°\text{C})^{-1}$

this becomes $\qquad\qquad\qquad L_S = \dfrac{5.00 \text{ cm}}{1 - 11/17} = 14.2 \text{ cm}$ ◊

Equation [2] then gives $\qquad\qquad L_C = L_S - 5.00 \text{ cm} = 9.2 \text{ cm}$ ◊

61. A bimetallic bar is made of two thin strips of dissimilar metals bonded together. As they are heated, the one with the larger average coefficient of expansion expands more than the other, forcing the bar into an arc, with the outer radius having a larger circumference (see Fig. P10.61). (a) Derive an expression for the angle of bending θ as a function of the initial length of the strips, their average coefficients of linear expansion, the change in temperature, and the separation of the centers of the strips ($\Delta r = r_2 - r_1$). (b) Show that the angle of bending goes to zero when ΔT goes to zero or when the two coefficients of expansion become equal. (c) What happens if the bar is cooled?

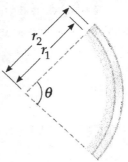

Figure P10.61

Solution

(a) Let L_i be the common initial length of the two strips.   After a change in temperature ΔT,

their lengths are $\qquad\qquad\qquad\qquad L_1 = L_i + \alpha_1 L_i(\Delta T)$

and $\qquad\qquad\qquad\qquad\qquad\qquad L_2 = L_i + \alpha_2 L_i(\Delta T)$

The lengths of the circular arcs are related to their radii by $L_1 = r_1\theta$ and $L_2 = r_2\theta$, where θ is measured in radians.

Thus, $\qquad\qquad\qquad\qquad \Delta r = r_2 - r_1 = \dfrac{L_2}{\theta} - \dfrac{L_1}{\theta} = \dfrac{(\alpha_2 - \alpha_1)L_i(\Delta T)}{\theta}$

or $\qquad\qquad\qquad\qquad\qquad \theta = \dfrac{(\alpha_2 - \alpha_1)L_i(\Delta T)}{\Delta r}$ $\qquad\qquad\qquad\qquad$ ◊

(b) As seen in the above result, $\qquad \theta = 0$ if either $\Delta T = 0$ or $\alpha_1 = \alpha_2$. $\qquad\qquad$ ◊

If $\Delta T < 0$, then θ is negative, meaning the bar bends in the direction opposite to that shown. $\qquad\qquad\qquad\qquad\qquad\qquad\qquad\qquad\qquad\qquad\qquad\qquad\qquad\qquad$ ◊

Chapter 11
ENERGY IN THERMAL PROCESSES

NOTES ON SELECTED CHAPTER SECTIONS

11.1   Heat and Internal Energy

When two systems at different temperatures are placed in thermal contact, energy is transferred by heat from the warmer to the cooler object until they reach a common temperature (that is, when they are in thermal equilibrium with each other). Heat is a mechanism by which energy (measured in calories or joules) is transferred between a system and its environment due to a temperature difference.

The **mechanical equivalent of heat**, first measured by Joule, is given by $1 \, cal = 4.186 \, J$. This is the definition of the calorie.

11.2   Specific Heat

The specific heat, c, of any substance is defined as the amount of energy required to increase the temperature of a unit mass of that substance by one Celsius degree. The units of specific heat are $J/kg \cdot °C$.

11.3   Calorimetry

Calorimetry is a procedure carried out in an isolated system (usually a substance of unknown specific heat and water) in which only the transfer of thermal energy occurs. A negligible amount of mechanical work is done in the process. The law of conservation of energy in a calorimeter requires that the energy that leaves the warmer substance (of unknown specific heat) equals the energy that enters the water.

11.4   Latent Heat and Phase Change

A substance usually undergoes a change in temperature when energy is transferred by heat between it and its surroundings. There are situations, however, in which the transfer of energy does not result in a change in temperature. This is the case whenever the substance undergoes a physical alteration from one form to another, referred to as a **phase change**. Some common phase changes are solid to liquid, liquid to gas, and a change in crystalline structure of a solid. Every phase change involves a change in internal energy without an accompanying change in temperature.

The **latent heat of fusion** is a parameter used to characterize a solid-to-liquid phase change; the **latent heat of vaporization** characterizes the liquid-to-gas phase change.

11.5   Energy Transfer by Thermal Conduction

11.6   Energy Transfer by Convection

11.7   Energy Transfer by Radiation

There are three basic processes of thermal energy transfer. These are (1) conduction, (2) convection, and (3) radiation.

Conduction is a energy transfer process which occurs in a substance when there is a **temperature gradient** across the substance. That is, conduction of energy occurs only when the temperature of the substance is **not** uniform. For example, if you position a metal rod with one end in a flame, energy will flow from the hot end to the colder end. The rate of flow of energy along the rod is proportional to the cross-sectional area of the rod, the temperature gradient, and k, the thermal conductivity of the material of which the rod is made.

When energy transfer occurs as the result of the motion of material, such as the mixing of hot and cold fluids, the process is referred to as **convection**. Convection heating is used in conventional hot-air and hot-water heating systems. Convection currents produce changes in weather conditions when warm and cold air masses mix in the atmosphere.

Energy transfer by **radiation** is the result of the continuous emission of electromagnetic radiation by all bodies.

EQUATIONS AND CONCEPTS

When two systems initially at different temperatures are placed in contact with each other, energy will be transferred from the system at higher temperature to the system at lower temperature until the two systems reach a common temperature (thermal equilibrium).

Comment on thermal energy.

The **mechanical equivalent of heat** was first measured by Joule. This is the **definition of the calorie as a general energy unit**.

$$1 \text{ cal} = 4.186 \text{ J} \tag{11.1}$$

The quantity of energy required to increase the temperature of a given mass by a specified amount varies from one substance to another. Every substance is characterized by a unique value of specific heat, c. In this and subsequent equations Q represents the quantity of energy transferred by heat between a system and its environment.

$$Q = mc\Delta T \tag{11.3}$$

The energy required to cause a quantity of substance of mass, m, to undergo a phase change depends on the value of the latent heat of the substance. The latent heat of fusion, L_f is used when the phase change is from solid to liquid (or liquid to solid). The latent heat of vaporization, L_v is used when the phase change is from liquid to gas (or gas to liquid). The positive or negative sign is chosen according to the direction of energy flow.

$$Q = \pm mL \tag{11.5}$$

There are three basic processes by which thermal energy is transferred. These are:

    (1) conduction,
    (2) convection, and
    (3) radiation.

This is the basic **law of thermal conduction**. The constant k is called the thermal conductivity and is characteristic of a particular material. Conduction of energy occurs only when there is a temperature gradient (or temperature difference between two points in the conducting material).

$$\mathscr{P} = k A \left(\frac{T_h - T_c}{L} \right) \qquad (11.6)$$

To calculate the rate of energy transfer through a **compound slab** a summation is made over all portions of the slab. For this calculation, T_h and T_c are the temperatures of the **outer extremities** of the slab. In engineering practice, the term L/k for a particular substance is referred to as the **R value** of the material.

$$\frac{Q}{\Delta t} = \frac{A(T_h - T_c)}{\sum_i L_i/k_i} \qquad (11.7)$$

The rate of energy transfer, $\mathscr{P}$, can be expressed in cal/s, Btu/h, or watts (where $1\,W = 1\,J/s$ and $1\,Btu/h = 0.293\,W$).

Comment on units

When energy transfer occurs as the result of the movement of a substance (usually a fluid) between points at different temperatures, the process is called convection.

Comment on convection

The rate of energy transfer by radiation (the power radiated) is given by Stefan's law. **The radiated power is proportional to the fourth power of the absolute temperature.**

$$\mathcal{P} = \sigma A e T^4 \qquad (11.9)$$

(Radiation process)

In Equation 11.9, σ is a parameter whose value is 5.6696×10^{-8} W/m^2, T is the absolute temperature in K, and e (the emissivity of the radiating body) can have a value between 0 and 1 depending on the nature of the surface.

Comment on Stefan's law.

An object at temperature T in surroundings at temperature T_0 will experience a net radiated $(T > T_0)$ or absorbed $(T < T_0)$ power. At thermal equilibrium $(T = T_0)$, an object radiates and absorbs energy at the same rate and the temperature of the object remains constant.

$$\mathcal{P}_{net} = \sigma A e\left(T^4 - T_0{}^4\right) \qquad (11.10)$$

SUGGESTIONS, SKILLS, AND STRATEGIES

If you are having difficulty with calorimetry problems, one or more of the following factors should be considered:

1. Be sure your units are consistent throughout. That is, if you are using specific heats measured in $J/kg \cdot {}^{\circ}C$, be sure that masses are in kilograms and temperatures are in Celsius units throughout.

2. Losses and gains in energy are found by using $Q = mc\Delta T$ only for those intervals in which no phase changes occur. Likewise, the equations $Q = \pm mL_f$ and $Q = \pm mL_v$ are to be used only when phase changes **are** taking place.

3. Often sign errors occur in Equation 11.4 ($Q_{cold} = -Q_{hot}$). Remember to include the negative sign in the equation, and remember that **ΔT is always the final temperature minus the initial temperature**.

REVIEW CHECKLIST

▷ Understand the important distinction between internal energy and heat.

▷ Define and discuss the calorie, Btu, specific heat, and latent heat. Convert between calories, Btus, and joules.

▷ Use equations for specific heat, latent heat, temperature change and energy gain (loss) to solve calorimetry problems.

▷ Discuss the possible mechanisms which can give rise to energy transfer between a system and its surroundings; that is, conduction, convection and radiation; and give a realistic example of each transfer mechanism.

▷ Apply the basic law of thermal conduction, and Stefan's law for energy transfer by radiation.

SOLUTIONS TO SELECTED END-OF-CHAPTER PROBLEMS

7. A 75.0-kg weight-watcher wishes to climb a mountain to work off the equivalent of a large piece of chocolate cake rated at 500 (food) Calories. How high must the person climb? [1 (food) Calorie $= 10^3$ calories]

Solution The chemical energy which must be converted into other forms during the climb is

$$Q = 500 \text{ Calories}\left(\frac{10^3 \text{ cal}}{1 \text{ Calorie}}\right) = 5.00 \times 10^5 \text{ cal}$$

Using the mechanical equivalent of heat to convert this to standard SI units of energy gives

$$Q = 5.00 \times 10^5 \text{ cal}\left(\frac{4.186 \text{ J}}{1 \text{ cal}}\right) = 2.09 \times 10^6 \text{ J}$$

If we assume that this all goes to increasing gravitational potential energy as the person climbs, then $Q = \Delta PE_g = mgh$. The required height of the climb is then

$$h = \frac{Q}{mg} = \frac{2.09 \times 10^6 \text{ J}}{(75.0 \text{ kg})(9.80 \text{ m}/\text{s}^2)} = 2.85 \times 10^3 \text{ m} = 2.85 \text{ km} \qquad \Diamond$$

In actual practice, much of the chemical energy would be dissipated by radiation and by evaporation of perspiration as the person climbs, so the required height would be considerably smaller than that calculated above.

===

11. A 200-g aluminum cup contains 800 g of water in thermal equilibrium with the cup at 80 °C. The combination of cup and water is cooled uniformly so that the temperature decreases by 1.5 °C per minute. At what rate is energy being removed? Express your answer in watts.

Solution Consider a 1.0-minute time interval. During this interval, the temperature of the system (consisting of the aluminum cup and the water in the cup) decreases by 1.5 °C. The energy removed by heat from the system is

$$Q_{total} = Q_{cup} + Q_{water} = m_{cup}c_{Al}\Delta T_{cup} + m_{water}c_{water}\Delta T_{water}$$

The specific heats of aluminum and water are:

$$c_{Al} = 900 \text{ J / kg} \cdot {}^\circ\text{C} \quad \text{and} \quad c_{water} = 4186 \text{ J / kg} \cdot {}^\circ\text{C}$$

The energy removed by heat is thus:

$$Q_{total} = (0.200 \text{ kg})(900 \text{ J / kg} \cdot {}^\circ\text{C})(1.5\ {}^\circ\text{C}) + (0.800 \text{ kg})(4186 \text{ J / kg} \cdot {}^\circ\text{C})(1.5\ {}^\circ\text{C})$$

$$Q_{total} = 5.3 \times 10^3 \text{ J}$$

This happens over 60 seconds. Thus, energy is removed at a rate of

$$\mathcal{P} = Q_{total} / t = (5300 \text{ J} / 60 \text{ s}) = 88 \text{ J / s} = 88 \text{ W} \qquad \Diamond$$

15. An aluminum cup contains 225 g of water and a 40-g copper stirrer, all at 27 °C. A 400-g sample of silver at an initial temperature of 87 °C is placed in the water. The stirrer is used to stir the mixture gently until it reaches its final equilibrium temperature of 32 °C. Calculate the mass of the aluminum cup.

Solution

Assuming that the system consisting of the cup, stirrer, water, and silver sample is isolated from the environment, the total energy content of the system is constant. Therefore, any energy gained by the initially cooler parts (cup, stirrer, and water) must equal the energy lost by the initially warmer parts (the silver sample).

Thus, $$Q_{cup} + Q_{stirrer} + Q_{water} = \left| Q_{Ag} \right|$$

Since the cup, stirrer, and water all undergo the same change in temperature, this becomes

$$\left[m_{cup} c_{Al} + m_{stirrer} c_{Cu} + m_{water} c_{water} \right] (\Delta T_{water}) = m_{Ag} c_{Ag} \left| \Delta T_{Ag} \right|$$

The mass of the cup is then

$$m_{cup} = m_{Ag} \left(\frac{c_{Ag}}{c_{Al}} \right) \left(\frac{\left| \Delta T_{Ag} \right|}{\Delta T_{water}} \right) - m_{stirrer} \left(\frac{c_{Cu}}{c_{Al}} \right) - m_{water} \left(\frac{c_{water}}{c_{Al}} \right)$$

or $$m_{cup} = (400 \text{ g}) \left(\frac{234}{900} \right) \left(\frac{87 - 32}{32 - 27} \right) - (40 \text{ g}) \left(\frac{387}{900} \right) - (225 \text{ g}) \left(\frac{4186}{900} \right) = 80 \text{ g} \qquad \Diamond$$

19. A student drops two metallic objects into a 120-g steel container holding 150 g of water at 25 °C. One object is a 200-g cube of copper that is initially at 85 °C, and the other is a chunk of aluminum that is initially at 5.0 °C. To the student's surprise, the water reaches a final temperature of 25 °C, precisely where it started. What is the mass of the aluminum chunk?

Solution   Notice that the initial temperature of the copper cube is higher than the final temperature. Thus, the copper cube loses energy by heat in this process. The initial temperature of the aluminum chunk is less than the final temperature, so it gains internal energy. The temperature of the water and the steel container is unchanged so these parts of the system have zero net change in internal energy. Thus, assuming the system is isolated from its surroundings, the energy balance equation is

$$Q_{Al} = |Q_{Cu}| \qquad \text{or} \qquad m_{Al}c_{Al}(\Delta T_{Al}) = m_{Cu}c_{Cu}|\Delta T_{Cu}|$$

The specific heats of aluminum and copper are

$$c_{Al} = 900 \text{ J / kg} \cdot °C \qquad \text{and} \qquad c_{Cu} = 387 \text{ J / kg} \cdot °C$$

so the mass of the aluminum chunk is

$$m_{Al} = m_{Cu}\left(\frac{c_{Cu}}{c_{Al}}\right)\left(\frac{|\Delta T_{Cu}|}{\Delta T_{Al}}\right) = (200 \text{ g})\left(\frac{387}{900}\right)\left(\frac{85-25}{25-5.0}\right) = 260 \text{ g} = 0.26 \text{ kg} \qquad \diamond$$

23. What mass of steam that is initially at 120 °C is needed to warm 350 g of water and its 300-g aluminum container from 20 °C to 50 °C?

Solution   Assuming the system consisting of the steam, water, and container is isolated from the environment, the energy content of the system must be constant. Therefore, the total energy lost by the steam must equal the total energy gained by the water and the container:

$$|Q_{steam}| = Q_{water} + Q_{container}$$

Since the final temperature of the system is below the boiling point of water, the steam must undergo a phase change as it cools. It must first cool to the boiling point at 100 °C, then condense into a liquid, and finally cool as condensed water to the final temperature at 50 °C. Thus,

$$|Q_{steam}| = m_{steam}c_{steam}(120 \text{ °C} - 100 \text{ °C}) + m_{steam}L_v + m_{steam}c_{water}(100 \text{ °C} - 50 \text{ °C})$$

where L_v is the latent heat of vaporization of water.

Using Tables 11.1 and 11.2 in the textbook, this becomes

$$|Q_{steam}| = m_{steam}\left[\left(2010\ \frac{J}{kg\cdot{}^{\circ}C}\right)(20\ {}^{\circ}C) + 2.26\times10^{6}\ J/kg + \left(4186\ \frac{J}{kg\cdot{}^{\circ}C}\right)(50\ {}^{\circ}C)\right]$$

or $\quad |Q_{steam}| = m_{steam}\left(2.5\times10^{6}\ J/kg\right)$

The total energy gained by the water and container is

$$Q_{water} + Q_{container} = \left[m_{water}c_{water} + m_{container}c_{Al}\right]\left(\Delta T_{water}\right)$$

$$= \left[(0.350\ kg)\left(4186\ \frac{J}{kg\cdot{}^{\circ}C}\right) + (0.300\ kg)\left(900\ \frac{J}{kg\cdot{}^{\circ}C}\right)\right](30\ {}^{\circ}C)$$

or $\quad Q_{water} + Q_{container} = 5.2\times10^{4}\ J$

The energy balance equation $\left(|Q_{steam}| = Q_{water} + Q_{container}\right)$ then yields

$$m_{steam} = \frac{5.2\times10^{4}\ J}{2.5\times10^{6}\ J/kg} = 2.1\times10^{-2}\ kg = 21\ g \qquad \Diamond$$

27. A 40-g block of ice is cooled to $-78\ {}^{\circ}C$. It is added to 560 g of water in an 80-g copper calorimeter at a temperature of $25\ {}^{\circ}C$. Determine the final temperature. (If not all the ice melts, determine how much ice is left.) Remember that the ice must first warm to $0\ {}^{\circ}C$, melt, and then continue warming as water. The specific heat of ice is $0.500\ cal/g\cdot{}^{\circ}C = 2090\ J/kg\cdot{}^{\circ}C$.

Solution   There are 3 possible outcomes of the process that must be considered. These are:

(1) The water and cup could reach $0\ {}^{\circ}C$ and all the water could freeze before the block of ice warms to $0\ {}^{\circ}C$. In this case, the equilibrium temperature T_f would be below $0\ {}^{\circ}C$.

(2) The water and cup may reach $0\ {}^{\circ}C$ after the block of ice has reached $0\ {}^{\circ}C$ but before all the ice has melted. Then, the equilibrium temperature will be $T_f = 0\ {}^{\circ}C$.

(3) The block of ice could reach $0\ {}^{\circ}C$ and completely melt before the water and cup have cooled to $0\ {}^{\circ}C$. In this case, $T_f > 0\ {}^{\circ}C$.

To distinguish between these possible scenarios, consider the quantities of energy that would be transferred in various steps. The internal energy the ice must gain before its temperature will reach $0\,°C$ is

$$Q_{\text{warm ice to } 0\,°C} = Q_1 = m_{\text{ice}}\,c_{\text{ice}}\left(0\,°C - T_{\text{ice, }i}\right)$$

or $\quad Q_1 = (0.040\text{ kg})(2090\text{ J / kg}\cdot°C)(78\,°C) = 6.5\times10^3\text{ J}$

After its temperature has reached $0\,°C$, the additional internal energy needed to completely melt the ice (with L_f being the heat of fusion for water) is:

$$Q_{\text{melt ice}} = Q_2 = m_{\text{ice}}L_f$$

or $\quad Q_2 = (0.040\text{ kg})(3.33\times10^5\text{ J / kg}) = 1.3\times10^4\text{ J}$

The internal energy the water and cup could lose before their temperature reaches $0\,°C$ is

$$Q_3 = \left(m_{\text{water}}c_{\text{water}} + m_{\text{cup}}c_{\text{cup}}\right)\left(T_{\text{water+cup, }i} - 0\,°C\right)$$

$$Q_3 = \left[(0.56\text{ kg})(4186\text{ J / kg}\cdot°C) + (0.080\text{ kg})(387\text{ J / kg}\cdot°C)\right](25\,°C) = 5.9\times10^4\text{ J}$$

Observe that $Q_3 > Q_1 + Q_2$. Thus, the water and cup are capable of providing more than enough energy to warm the ice to $0\,°C$ and completely melt the ice. Therefore, the equilibrium temperature is greater than $0\,°C$. This temperature can be determined from an energy balance equation for the isolated system: $Q_{\text{gain}} = Q_{\text{loss}}$.

Specifically, that is $\quad \underset{\text{to } 0\,°C}{Q_{\text{warm ice}}} + \underset{\text{ice}}{Q_{\text{melt}}} + \underset{\text{melted ice}}{Q_{\text{warm}}} = \underset{\text{plus cup}}{Q_{\text{cool water}}}$

Recognize that after the ice melts, its specific heat is simply that of liquid water. Then, the energy balance equation becomes

$$Q_1 + Q_2 + m_{\text{ice}}c_{\text{water}}\left(T_f - 0\,°C\right) = \left(m_{\text{water}}c_{\text{water}} + m_{\text{cup}}c_{\text{cup}}\right)\left(25\,°C - T_f\right)$$

$$6.5\times10^3\text{ J} + 1.3\times10^4\text{ J} + (0.040\text{ kg})(4186\text{ J/kg}\cdot°C)T_f =$$

$$\left[(0.56\text{ kg})(4186\text{ J/kg}\cdot°C) + (0.080\text{ kg})(387\text{ J/kg}\cdot°C)\right]\left(25\,°C - T_f\right)$$

This yields $2.0\times10^4\text{ J} + \left(1.7\times10^2\text{ J / }°C\right)T_f = 6.0\times10^4\text{ J} - \left(2.4\times10^3\text{ J / }°C\right)T_f$

and $\quad T_f = \dfrac{6.0\times10^4\text{ J} - 2.0\times10^4\text{ J}}{1.7\times10^2\text{ J / }°C + 2.4\times10^3\text{ J / }°C} = 16\,°C$ $\qquad\qquad \diamond$

38. A Thermopane window consists of two glass panes, each 0.50 cm thick, with a 1.0-cm-thick sealed layer of air between. If the inside surface temperature is 23 °C and the outside surface temperature is 0.0 °C, determine the rate of energy transfer through 1.0 m^2 of the window. Compare this with the rate of energy transfer through 1.0 m^2 of a single 1.0-cm-thick pane of glass.

Solution

The rate of energy transfer through a slab which is made of multiple layers of materials may be written as $Q / \Delta t = A(T_h - T_c) / \Sigma R_i$, where A is the surface area of one side of the slab, $(T_h - T_c)$ is the total change in temperature going from one side of the slab to the other, and R_i is the R value for the ith layer making up the slab. The R value for any given layer is the ratio of its thickness L to the thermal conductivity, k, of the material making up the layer. The thermal conductivities of glass and air are 0.84 J / s·m·°C and 0.0234 J / s·m·°C, respectively. For the Thermopane, the R values for the various layers are

$$R_{\substack{glass \\ pane}} = \frac{5.0 \times 10^{-3} \text{ m}}{0.84 \text{ J / s·m·°C}} = 0.0060 \text{ s·m}^2 \text{·°C / J}$$

$$R_{\substack{air \\ layer}} = \frac{1.0 \times 10^{-2} \text{ m}}{0.0234 \text{ J / s·m·°C}} = 0.43 \text{ s·m}^2 \text{·°C / J}$$

Thus, with a temperature change of $(T_h - T_c) = 23 \text{ °C} - 0 \text{ °C}$ across it and a surface area $A = 1.0 \text{ m}^2$, the rate of energy transfer through the Thermopane is

$$\left(\frac{Q}{\Delta t}\right)_{Thermopane} = \frac{(1.0 \text{ m}^2)(23 \text{ °C} - 0 \text{ °C})}{(6.0 \times 10^{-3} + 0.43 + 6.0 \times 10^{-3}) \text{ s·m}^2 \text{·°C / J}} = 52 \text{ J / s} = 52 \text{ W} \qquad \Diamond$$

For the single pane window, the R value of the thick glass pane is

$$R_{thick \ pane} = 1.0 \times 10^{-2} \text{ m} / (0.84 \text{ J/s·m·°C}) = 1.2 \times 10^{-2} \text{ s·m}^2 \text{·°C/J}$$

Then, for the same area and temperature difference as the Thermopane, the rate of energy transfer for the single pane is

$$(Q/\Delta t)_{\substack{single \\ pane}} = \frac{(1.0 \text{ m}^2)(23 \text{ °C} - 0 \text{ °C})}{1.2 \times 10^{-2} \text{ J / s·m·°C}} = 1.9 \times 10^3 \text{ J / s} = 1.9 \text{ kW} \qquad \Diamond$$

41. A sphere that is a perfect black-body radiator has a radius of 0.060 m and is at 200 °C in a room where the temperature is 22 °C. Calculate the net rate at which the sphere radiates energy.

Solution The rate at which a body of surface area A and **absolute temperature** T radiates energy is given by Stefan's law as

$$P_{radiation} = \sigma A e T^4$$

where e is the emissivity of that body and $\sigma = 5.6696 \times 10^{-8}$ W/m$^2 \cdot$K^4 is a constant.

If the temperature of the surroundings of that body is T_0 the body also **absorbs** energy (i.e., has a negative flow of energy from the body) at a rate

$$P_{absorption} = \sigma A e T_0^4$$

The **net** rate of energy transfer **away** from the body is then given by

$$P_{net} = P_{radiation} - P_{absorption} = \sigma A e \left(T^4 - T_0^4\right)$$

The surface area of a sphere of radius R is $A = 4\pi R^2$ and the emissivity of a perfect black-body is $e = 1.0$.

Thus, if a spherical black-body radiator with a radius of 0.060 m is at a temperature $T = 200$ °C $= 473$ K and the temperature of the surroundings is $T_0 = 22$ °C $= 295$ K, the net rate of radiation from the sphere is

$$P_{net} = \left(5.6696 \times 10^{-8} \text{ W/m}^2 \cdot \text{K}^4\right)\left[4\pi(0.060 \text{ m})^2\right](1.0)\left[(473 \text{ K})^4 - (295 \text{ K})^4\right]$$

or $P_{net} = 1.1 \times 10^2$ W $= 0.11$ kW   ◊

47. The bottom of a copper kettle has a 10-cm radius and is 2.0 mm thick. The temperature of the outside surface is 102 °C, and the water inside the kettle is boiling at 1 atm of pressure. Find the rate at which energy is being transferred through the bottom of the kettle.

Solution The rate of energy transfer by conduction through a slab of material having thickness L is

$$P = \frac{Q}{\Delta t} = kA\left(\frac{\Delta T}{L}\right)$$

Here, k is the thermal conductivity of the material, A is the surface area of one face of the slab, and ΔT is the difference in the temperatures of the two faces of the slab.

At a pressure of 1 atm, water boils at 100 °C. Thus, $\Delta T = 102\ °C - 100\ °C$ for the kettle bottom.

Also, $A = \pi r^2 = \pi(0.100\ \text{m})^2$ and $k = 397\ \text{J} / \text{s} \cdot \text{m} \cdot °C$ for copper.

The desired transfer rate is then

$$\mathcal{P} = (397\ \text{J} / \text{s} \cdot \text{m} \cdot °C)\left[\pi(0.10\ \text{m})^2\right]\left(\frac{102\ °C - 100\ °C}{2.00 \times 10^{-3}\ \text{m}}\right) = 1.2 \times 10^4\ \text{W} = 12\ \text{kW} \qquad \Diamond$$

---

51. A 40-g ice cube floats in 200 g of water in a 100-g copper cup; all are at a temperature of 0 °C. A piece of lead at 98 °C is dropped into the cup, and the final equilibrium temperature is 12 °C. What is the mass of the lead?

Solution This problem is easily solved by the methods of calorimetry (i.e., conservation of energy). However, it is necessary to remember that the ice must completely melt before its temperature (and hence that of the water and cup) will rise above 0 °C.

Also, after the ice melts, the cup will contain a total of 240 g liquid water. The energy balance equation $|Q_{\text{loss}}| = Q_{\text{gain}}$ for this isolated system becomes:

$$|Q_{\text{cool lead}}| = Q_{\text{melt ice}} + Q_{\substack{\text{warm water} \\ \text{and melted ice}}} + Q_{\text{warm cup}}$$

or $\quad m_{\text{lead}}c_{\text{lead}}\left(T_{\text{lead},\,i} - T_f\right) = m_{\text{ice}}L_f + m_{\substack{\text{water} \\ +\ \text{ice}}} c_{\text{water}}\left(T_f - 0\ °C\right) + m_{\text{cup}}c_{\text{Cu}}\left(T_f - 0\ °C\right)$

From Tables 11.1 and 11.2, $\qquad c_{\text{lead}} = 128\ \text{J} / \text{kg} \cdot °C \qquad c_{\text{Cu}} = 387\ \text{J} / \text{kg} \cdot °C$

$$c_{\text{water}} = 4186\ \text{J} / \text{kg} \cdot °C \qquad \left(L_f\right)_{\text{water}} = 3.33 \times 10^5\ \text{J} / \text{kg}$$

Thus, $\quad m_{\text{lead}}(128\ \text{J} / \text{kg} \cdot °C)(98\ °C - 12\ °C) = (0.040\ \text{kg})\left(3.33 \times 10^5\ \text{J} / \text{kg}\right) +$

$$(0.240\ \text{kg})\left(4186\ \text{J} / \text{kg} \cdot °C\right)(12\ °C) +$$

$$(0.100\ \text{kg})\left(387\ \text{J} / \text{kg} \cdot °C\right)(12\ °C)$$

or $\quad m_{\text{lead}} = \dfrac{2.58 \times 10^4\ \text{J}}{1.1 \times 10^4\ \text{J} / \text{kg}} = 2.3\ \text{kg}$ $\qquad \Diamond$

---

56. An aluminum rod and an iron rod are joined end to end in a good thermal contact. The two rods have equal lengths and radii. The free end of the aluminum rod is maintained at a temperature of 100 °C, and the free end of the iron rod is maintained at 0 °C. (a) Determine the temperature of the interface between the two rods. (b) If each rod is 15 cm long and each has a cross-sectional area of 5.0 cm², what quantity of energy is conducted across the combination in 30 min?

Solution

(a) At steady state, the rate of transfer of energy to the interface through the aluminum rod equals the rate of transfer of energy away from the interface through the iron rod. That is:

$$P = \frac{Q}{\Delta t} = \frac{k_{Al} A_{Al} (\Delta T)_{Al}}{L_{Al}} = \frac{k_{Fe} A_{Fe} (\Delta T)_{Fe}}{L_{Fe}}$$

where k_{Al} and k_{Fe} are the thermal conductivities of aluminum and iron, respectively. The rods have equal lengths $(L_{Al} = L_{Fe})$, and since they have the same radii, their cross-sectional areas are also equal (i.e., $A_{Al} = A_{Fe}$).

The energy transfer equation then reduces to $\quad k_{Al}(\Delta T)_{Al} = k_{Fe}(\Delta T)_{Fe}$

or $\qquad\qquad\qquad k_{Al}(T_1 - T_I) = k_{Fe}(T_I - T_2)$

From Table 11.3 in the textbook, $\qquad k_{Al} = 238 \text{ J}/\text{s}\cdot\text{m}\cdot°\text{C}$

and $\qquad\qquad\qquad k_{Fe} = 79.5 \text{ J}/\text{s}\cdot\text{m}\cdot°\text{C}$

Also, it is given that $\qquad\qquad T_1 = 100 \text{ °C} \quad\text{and}\quad T_2 = 0 \text{ °C}$

Thus, $(238 \text{ J}/\text{s}\cdot\text{m}\cdot°\text{C})(100 \text{ °C} - T_I) = (79.5 \text{ J}/\text{s}\cdot\text{m}\cdot°\text{C})(T_I - 0 \text{ °C})$

Solving for the temperature at the interface between the two rods gives

$$2.38 \times 10^4 \text{ J}/\text{s}\cdot\text{m} = (238 \text{ J}/\text{s}\cdot\text{m}\cdot°\text{C} + 79.5 \text{ J}/\text{s}\cdot\text{m}\cdot°\text{C})\, T_I \quad\text{and}\quad T_I = 75 \text{ °C} \;\Diamond$$

(b) If the aluminum rod has a length of $\qquad L_{Al} = 15 \text{ cm} = 0.15 \text{ m}$

and a cross-sectional area of $\qquad A_{Al} = 5.0 \text{ cm}^2 = 5.0 \times 10^{-4} \text{ m}^2$
the rate of transfer through that rod is

$$P = \frac{Q}{\Delta t} = \frac{k_{Al} A_{Al}(T_1 - T_I)}{L_{Al}} = \frac{(238 \text{ J}/\text{s}\cdot\text{m}\cdot°\text{C})(5.0 \times 10^{-4} \text{ m}^2)(25 \text{ °C})}{0.15 \text{ m}} = 20 \text{ J}/\text{s}$$

The energy transferred in 30 min is

$$Q = (20 \text{ J}/\text{s})\Delta t = (20 \text{ J}/\text{s})(30 \text{ min})(60 \text{ s}/\text{min}) = 3.6 \times 10^4 \text{ J} \qquad\qquad \Diamond$$

63. A **flow calorimeter** is an apparatus used to measure the specific heat of a liquid. The technique is to measure the temperature difference between the input and output points of a flowing stream of the liquid while adding energy at a known rate.

(a) Start with the equations $Q = mc(\Delta T)$ and $m = \rho V$ and show that the rate at which energy is added to the liquid is given by the expression

$$\frac{\Delta Q}{\Delta t} = \rho c (\Delta T) \left(\frac{\Delta V}{\Delta t} \right)$$

(b) In a particular experiment, a liquid of density $0.72 \text{ g} / \text{cm}^3$ flows through the calorimeter at the rate of $3.5 \text{ cm}^3/\text{s}$. At steady state, a temperature difference of 5.8 °C is established between the input and output points when energy is supplied at the rate of $40 \text{ J} / \text{s}$. What is the specific heat of the liquid?

Solution

(a) The energy ΔQ added to the volume ΔV of liquid that flows through the calorimeter in time Δt is

$$\Delta Q = (\Delta m) c (\Delta T) \quad [\rho(\Delta V)] c (\Delta T)$$

Thus, the rate of adding energy to the liquid is

$$\frac{\Delta Q}{\Delta t} = \rho c (\Delta T) \left(\frac{\Delta V}{\Delta t} \right) \qquad\qquad ◊$$

where $(\Delta V / \Delta t)$ is the flow rate through the calorimeter.

(b) From the result of part (a), the specific heat of the flowing liquid is

$$c = \frac{\Delta Q / \Delta t}{\rho(\Delta T)(\Delta V / \Delta t)} = \frac{40 \text{ J} / \text{s}}{(0.72 \text{ g} / \text{cm}^3)(5.8 \text{ °C})(3.5 \text{ cm}^3 / \text{s})}$$

or $\quad c = (2.7 \text{ J} / \text{g} \cdot \text{°C}) \left(\frac{10^3 \text{ g}}{1 \text{ kg}} \right) = 2.7 \times 10^3 \text{ J} / \text{kg} \cdot \text{°C} \qquad\qquad ◊$

Chapter 12

THE LAWS OF THERMODYNAMICS

NOTES ON SELECTED CHAPTER SECTIONS

12.1 Work in Thermodynamic Processes

The work done on a gas in a process that takes it from some initial state to some final state is the negative of the area under the curve on a PV diagram. (See the figure to the right.)

If the gas is compressed ($V_f < V_i$) the work done on the gas is positive. If the gas expands ($V_f > V_i$) the work done on the gas is negative. If the gas expands at **constant pressure**, called an **isobaric process**, then $W = -P(V_f - V_i)$.

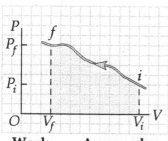

Work on gas = −Area under the curve

The work done on a system depends on the process by which the system goes from the initial to the final state. In other words, the work done depends on the initial, final, and intermediate states of the system.

12.2 The First Law of Thermodynamics

In the first law of thermodynamics, $\Delta U = Q + W$, Q is the energy transferred to the system by heat and W is the work done on the system. Note that by convention, Q is **positive** when energy enters the system and **negative** when energy is removed from the system. Likewise, W can be positive or negative as mentioned earlier. The initial and final states must be **equilibrium** states; however, the intermediate states are, in general, nonequilibrium states since the thermodynamic coordinates undergo finite changes during the thermodynamic process.

12.4 Heat Engines and the Second Law of Thermodynamics

A heat engine is a device that converts internal energy to other useful forms, such as electrical and mechanical energy. A heat engine carries some working substance through a cyclic process during which (1) energy is transferred from a reservoir at a high temperature, (2) work is done by the engine, and (3) energy is expelled by the engine to a reservoir at a lower temperature.

The engine absorbs a quantity of energy, Q_h, from a hot reservoir, does work W_{eng}, and then gives up energy Q_c to a cold reservoir. Because the working substance goes through a cycle, its initial and final internal energies are equal, so $\Delta U = 0$. Hence, from the first law of thermodynamics, $\Delta U = Q + W$, we see that **the work done by a heat engine equals the energy absorbed from the reservoirs:** $W_{eng} = Q_h - Q_c$.

If the working substance is a gas, **the work done by the engine for a cyclic process is the area enclosed by the curve representing the process on a PV diagram.**

The **thermal efficiency**, e, of a heat engine is the ratio of the work done by the engine to the energy absorbed at the higher temperature during one cycle.

The second law of thermodynamics can be stated as follows: **It is impossible to construct a heat engine that, operating in a cycle, produces no other effect than the absorption of heat from a reservoir and the performance of an equal amount of work.**

12.5   Reversible and Irreversible Processes

A process is **irreversible** if the system and its surroundings cannot be returned to their initial states. A process is **reversible** if the system passes from the initial to the final state through a succession of equilibrium states and can be returned to its initial condition along the same path.

12.6   The Carnot Engine

The **Carnot cycle** is the most efficient cyclic process (or engine) operating between two given energy reservoirs with temperatures T_h and T_c. The PV diagram for the Carnot cycle, illustrated in the figure, includes four reversible processes: two isothermal and two adiabatic.

1. The process $A \rightarrow B$ is an isotherm (constant T), during which time the gas expands at constant temperature T_h and absorbs energy Q_h from the hot reservoir.

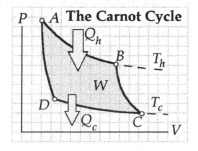

2. The process $B \rightarrow C$ is an adiabatic expansion ($Q = 0$), during which time the gas expands and cools to a temperature T_c.

185

3. The process $C \rightarrow D$ is a second isotherm, during which time the gas is compressed at constant temperature T_c, and expels energy Q_c to the cold reservoir.

4. The final process $D \rightarrow A$ is an adiabatic compression in which the gas temperature increases to a final temperature of T_h.

In practice, no working engine is 100% efficient, even when losses such as friction are neglected. One can obtain some theoretical limits on the efficiency of a real engine by comparison with the ideal Carnot engine. A **reversible engine** is one which will operate with the same efficiency in the forward and reverse directions. The Carnot engine is one example of a reversible engine.

All Carnot engines operating reversibly between T_h and T_c have the **same** efficiency given by Equation 12.8.

No real (irreversible) engine can have an efficiency greater than that of a reversible engine operating between the same two temperatures.

12.7 Entropy

12.8 Entropy and Disorder

Entropy is a quantity used to measure the degree of **disorder** in a system. For example, the molecules of a gas in a container at a high temperature are in a more disordered state (higher entropy) than the same molecules at a lower temperature.

When energy is transferred to a system by heat, the entropy **increases**. When energy is transferred out of a system by heat, the entropy **decreases**. In describing a thermodynamic process, the **change in entropy is the important quantity**; therefore the concept of entropy is most useful when a system undergoes a **change in its state**.

The second law of thermodynamics can be stated in terms of entropy as follows: **The total entropy of an isolated system always increases in time if the system undergoes an irreversible process**. If an isolated system undergoes a **reversible** process, the total entropy **remains constant**.

186

EQUATIONS AND CONCEPTS

This equation can be used to calculate the work **done on** a gas sample, if the pressure of the gas remains constant during a compression or an expansion. When ΔV is negative (compression), the work done on the gas is positive; when ΔV is positive (expansion), the work done on the gas is negative. The work done is equal to the negative of the area under the pressure-volume curve.

$$W = -P\Delta V \tag{12.1}$$

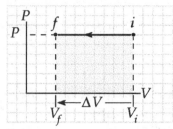

This is a statement of the **first law of thermodynamics** in equation form. This law is a special case of the general principle of conservation of energy and includes changes in the internal energy of a system. The change in the internal energy of a system is equal to the sum of the energy transferred across the boundary by heat and the energy transferred by work.

$$\Delta U = Q + W \tag{12.2}$$

Q = energy added to system by heat
W = work done on system
ΔU = change in the internal energy of system

Q is positive when energy is added by heat to the system. W is positive when work is done on the system by its surroundings.

Comment on the
first law of thermodynamics

The values of both Q and W depend on the path or sequence of processes by which a system changes from an initial to a final state. The internal energy U depends only on the initial and final states of the system.

The following definitions will be important in describing some applications of the laws of thermodynamics.

A system which does not interact with its surroundings.

Isolated system

In such a system $Q = W = 0$, so that $\Delta U = 0$. The internal energy of an isolated system remains constant.

A process for which the initial and final states are the same.

Cyclic process

For a cyclic process, $\Delta U = 0$ and $Q = -W$. The net work done per cycle by the gas equals the area enclosed by the path representing the process on a PV diagram.

A process in which no energy is transferred by heat.

Adiabatic process

$Q = 0$ and, therefore, the change in internal energy equals the work done on the system $(\Delta U = W)$. A system can undergo an adiabatic process if it is thermally insulated from its surroundings.

A process which occurs at constant pressure.

Isobaric process

In an isobaric process, the work done and the energy transferred by heat are both nonzero.

A process which occurs at constant volume.

Isovolumentric process

Since at constant volume, $\Delta V = 0$, $W = 0$; and therefore, $\Delta U = Q$. The net energy added by heat at constant volume goes into increasing the internal energy.

A process which occurs at constant temperature.

Isothermal process

When a monoatomic ideal gas undergoes an isothermal process, $\Delta U = 0$ and $W = -Q$ (the work done on the system is equal to the negative of the energy added by heat.)

A device that converts thermal energy into other forms of useful energy by carrying a substance through a cycle:

Heat Engine

(1) energy is transferred from a reservoir at a high temperature,

(2) work is done by the engine,

(3) energy is expelled by the engine to a reservoir at a low temperature.

A measure of the degree of disorder in a system.

Entropy

The net work, W_{eng}, done by a heat engine equals the net energy absorbed by the engine. Q_h is the quantity of energy absorbed from the high temperature reservoir and Q_c is the quantity of energy expelled to the low temperature reservoir.

$$W_{eng} = |Q_h| - |Q_c| \qquad (12.6)$$

The thermal efficiency of a heat engine is the ratio of the work done to the energy absorbed at the higher temperature during one cycle of the process.

$$e = \frac{|Q_h| - |Q_c|}{|Q_h|} = 1 - \frac{|Q_c|}{|Q_h|} \qquad (12.7)$$

It is impossible to construct a heat engine that, operating in a cycle, produces no other effect than the absorption of energy from a reservoir and the performance of an equal amount of work.

Kelvin-Planck statement of the second law of thermodynamics.

The Carnot cycle is the most efficient cyclic process and the thermal efficiency of an ideal Carnot engine depends on the temperatures of the hot and cold reservoirs. All real engines are less efficient than the Carnot engine.

$$e_c = \frac{T_h - T_c}{T_h} = 1 - \frac{T_c}{T_h} \qquad (12.8)$$

Entropy, S, is a thermodynamic variable which characterizes the degree of disorder in a system. All physical processes tend toward a state of increasing entropy. In going from an initial to a final state, the **change in entropy**, ΔS, is the ratio of the energy transferred to the system along a reversible path, to the absolute temperature of the system.

$$\Delta S \equiv \frac{Q_r}{T} \qquad\qquad (12.9)$$

The entropy of the Universe increases in all natural processes. This is an alternate way of expressing the second law of thermodynamics.

REVIEW CHECKLIST

▷ Understand how work is defined when a system undergoes a change in state, and the fact that work depends on the path taken by the system. You should also know how to sketch processes on a PV diagram, and calculate work using these diagrams.

▷ State the first law of thermodynamics $(\Delta U = Q + W)$, and explain the meaning of the three forms of energy contained in this statement. Discuss the implications of the first law of thermodynamics as applied to (i) an isolated system, (ii) a cyclic process, (iii) an adiabatic process, and (iv) an isothermal process.

▷ Describe the processes via which an ideal heat engine goes through a **Carnot cycle**. Express the efficiency of an ideal heat engine (Carnot engine) as a function of work and energy exchange with its environment. Express the maximum efficiency of an ideal heat engine as a function of its input and output temperatures.

▷ Understand the concept of entropy. Define **change in entropy** for a system in terms of its energy gain or loss by heat, and its temperature. State the **second law of thermodynamics** as it applies to entropy changes in a thermodynamic system.

SOLUTIONS TO SELECTED END-OF-CHAPTER PROBLEMS

1. The only form of energy possessed by molecules of a monatomic ideal gas is translational kinetic energy. Using the results from the discussion of kinetic theory in Section 10.6, show that the internal energy of a monatomic ideal gas at pressure P and occupying volume V may be written as $U = \frac{3}{2}PV$.

Solution

The average kinetic energy per molecule in a monatomic ideal gas is

$$KE_{molecule} = \frac{3}{2}k_B T$$

where T is the absolute temperature of the gas and the Boltzmann constant is

$$k_B = \frac{\text{universal gas constant}}{\text{Avogadro's number}} = \frac{R}{N_A}$$

If the gas contains N molecules, the total kinetic energy associated with random thermal motions is

$$KE = N\left(KE_{molecule}\right) = \frac{3}{2}N\left(\frac{R}{N_A}\right)T = \frac{3}{2}\left(\frac{N}{N_A}\right)RT$$

The total number of molecules, N, divided by the number of molecules in a mole, N_A, gives the number of moles of gas present, n.

Thus, the total kinetic energy becomes $KE = \frac{3}{2}nRT$

In an ideal gas, there are no intermolecular forces, so there are no potential energies contributing to the internal energy of the gas. The internal energy is therefore the same as the total kinetic energy,

or $$U = \frac{3}{2}nRT$$

Making use of the ideal gas law, $PV = nRT$

this internal energy may be written as $U = \frac{3}{2}PV$ ◊

5. A gas expands from I to F along the three paths indicated in Figure P12.5. Calculate the work done **on** the gas along paths (a) IAF, (b) IF, and (c) IBF.

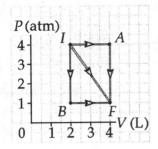

Figure P12.5

Solution In any process, the work done on the gas is given by the negative of the area under the path on the PV diagram. Along those parts of the path where volume is constant, no work is done.

(a) The area under the path IAF in the PV diagram of Figure P12.5 is a rectangle

of height
$$P_I = P_A = 4.00 \text{ atm}$$

and width
$$V_A - V_I = (4.00 - 2.00) \text{ liters} = 2.00 \text{ liters}$$

The work done **on** the gas is then

$$W_{IAF} = -(4.00 \text{ atm})(2.00 \text{ liters}) = -8.00 \text{ atm} \cdot \text{liter}$$

which can be converted to standard SI units by the following conversion factor:

$$1 \text{ atm} \cdot \text{liter} = (1 \text{ atm} \cdot \text{liter})\left(\frac{1.013 \times 10^5 \text{ N} / \text{m}^2}{1 \text{ atm}}\right)\left(\frac{10^{-3} \text{ m}^3}{1 \text{ liter}}\right) = 101.3 \text{ J}$$

Thus, $W_{IAF} = -8.00 \text{ atm} \cdot \text{liter}(101.3 \text{ J/atm} \cdot \text{liter}) = -810 \text{ J}$ ◊

(b) The area under the path IF on the PV diagram consists of a triangle on top of a rectangle. The triangle has a height $P_I - P_B = 3.00 \text{ atm}$ and the height of the rectangle is $P_B = 1.00 \text{ atm}$. The base of the triangle and the width of the rectangle both equal $V_F - V_B = 2.00 \text{ liters}$. Thus, the work done on the gas is

$$W_{IF} = -\frac{1}{2}(2.00 \text{ liters})(3.00 \text{ atm}) - (1.00 \text{ atm})(2.00 \text{ liters})$$

or $\quad W_{IF} = -5.00 \text{ atm} \cdot \text{liter} = -5.00 \text{ atm} \cdot \text{liter}\left(\frac{101.3 \text{ J}}{1 \text{ atm} \cdot \text{liter}}\right) = -507 \text{ J}$ ◊

(c) The area under the path IBF is a rectangle of height $P_B = 1.00 \text{ atm}$ and width $V_F - V_B = 2.00 \text{ liters}$. The work done on the gas along this path is

$$W_{IBF} = -(1.00 \text{ atm})(2.00 \text{ liters}) = -2.00 \text{ atm} \cdot \text{liter}\left(\frac{101.3 \text{ J}}{1 \text{ atm} \cdot \text{liter}}\right) = -203 \text{ J} \quad ◊$$

12. A quantity of a monatomic ideal gas undergoes a process in which both its pressure and volume are doubled as shown in Figure P12.12. What is the energy absorbed by heat into the gas during this process? (**Hint:** See Problem 1 above.)

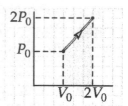

Solution The work done on the gas is equal to the negative of the shaded area under the process curve on the PV diagram. This is a triangle on top of a rectangle, so

Figure P12.12 (modified)

$$W = -\tfrac{1}{2}(2P_0 - P_0)(2V_0 - V_0) - P_0(2V_0 - V_0) = -1.5P_0V_0$$

Using Problem 12.1, the change in the internal energy of the gas is seen to be

$$\Delta U = U_f - U_0 = \tfrac{3}{2}P_fV_f - \tfrac{3}{2}P_0V_0 = \tfrac{3}{2}\left[(2P_0)(2V_0) - P_0V_0\right] = 4.5P_0V_0$$

The first law of thermodynamics, $\Delta U = Q + W$, then gives the energy absorbed by heat into the gas as

$$Q = \Delta U - W = 4.5P_0V_0 - (-1.5P_0V_0) = 6P_0V_0 \qquad \Diamond$$

---

15. A gas expands from I to F in Figure P12.5. The energy added to the gas by heat is 418 J when the gas goes from I to F along the diagonal path. (a) What is the change in internal energy of the gas? (b) How much energy must be added to the gas by heat for the indirect path IAF to give the same change in internal energy?

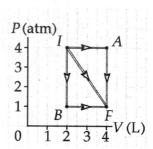

Solution

(a) The area under the diagonal path IF is a triangle, of height 3.00 atm, sitting atop a rectangle with a height of 1.00 atm. The base of the triangle is the same as the base of the rectangle, 2.00 L. The work done on the gas during this process is then

Figure P12.5

$$W = -(\text{area under curve}) = -\tfrac{1}{2}(3.00 \text{ atm})(2.00 \text{ L}) - (1.00 \text{ atm})(2.00 \text{ L})$$

or $$W = -5.00 \text{ atm} \cdot \text{L}\left(\frac{1.013 \times 10^5 \text{ N / m}^2}{1 \text{ atm}}\right)\left(\frac{10^{-3} \text{ m}^3}{1 \text{ L}}\right) = -506.5 \text{ J}$$

Since the energy added to the gas by heat is $Q = 418$ J for this process, the first law of thermodynamics gives the change in the internal energy as

$$U_F - U_I = \Delta U = Q + W = 418 \text{ J} - 506.5 \text{ J} = -88.5 \text{ J} \qquad \Diamond$$

(b)   The area under the indirect path IAF is a simple rectangle, of height 4.00 atm and width 2.00 L. The work done on the gas during this process is

$$W = -(\text{area under curve}) = -(4.00 \text{ atm})(2.00 \text{ L}) = -8.00 \text{ atm} \cdot \text{L}$$

or $$W = -8.00 \text{ atm} \cdot \text{L} \left(\frac{1.013 \times 10^5 \text{ N} / \text{m}^2}{1 \text{ atm}} \right) \left(\frac{10^{-3} \text{ m}^3}{1 \text{ L}} \right) = -810 \text{ J}$$

The change in the internal energy of the gas between states F and I is the same as found above, $\Delta U = U_F - U_I = -88.5 \text{ J}$. The first law of thermodynamics, $\Delta U = Q + W$, then gives the energy added to the gas by heat in the process IAF as

$$Q = \Delta U - W = -88.5 \text{ J} - (-810 \text{ J}) = 722 \text{ J} \qquad \Diamond$$

---

19.   One gram of water changes from liquid to solid at a constant pressure of one 1.00 atm and a constant temperature of 0 °C. In the process, the volume changes from 1.000 cm^3 to 1.090 cm^3. (a) Find the work done **on** the water, and (b) the change in the internal energy of the water.

Solution

(a)   In a constant pressure process, the work done on a thermodynamic system by its surroundings is $W = -P(\Delta V)$. In the process described, the pressure is

$$P = 1.00 \text{ atm} = 1.013 \times 10^5 \text{ Pa}$$

and $$\Delta V = V_f - V_i = (0.090 \text{ cm}^3) \left(\frac{1.00 \text{ m}^3}{1.00 \times 10^6 \text{ cm}^3} \right) = 9.00 \times 10^{-8} \text{ m}^3$$

The work done on the water is then

$$W = -(1.013 \times 10^5 \text{ Pa})(9.00 \times 10^{-8} \text{ m}^3) = -9.12 \times 10^{-3} \text{ J} \qquad \Diamond$$

(b)   From the first law of thermodynamics, the change in internal energy of the system is $\Delta U = Q + W$, where Q is the energy added to the system by heat during the process. In this case, energy in an amount equal to the heat of fusion must be **removed** (i.e., $Q < 0$) from the system to convert liquid water at 0 °C into a solid at 0 °C.

Thus, $$Q = -mL_f = -(1.00 \times 10^{-3} \text{ kg})(3.33 \times 10^5 \text{ J/kg}) = -333 \text{ J}$$

and $$\Delta U = -333 \text{ J} - 9.12 \times 10^{-3} \text{ J} = -333 \text{ J} \qquad \Diamond$$

27. One of the most efficient engines ever built is a coal-fired steam turbine in the Ohio valley, driving an electric generator as it operates between 1870 °C and 430 °C. (a) What is its maximum theoretical efficiency? (b) Its actual efficiency is 42.0%. How much mechanical power does the engine deliver if it absorbs 1.40×10^5 J of energy each second from the hot reservoir?

Solution

(a)   The maximum theoretical efficiency is that of a Carnot engine operating between the specified reservoirs. The **absolute** temperatures of the given hot and cold reservoirs are:

$$T_h = 1870 + 273 = 2143 \text{ K} \qquad \text{and} \qquad T_c = 430 + 273 = 703 \text{ K}$$

The Carnot efficiency is then

$$e_C = \frac{T_h - T_c}{T_h} = 1 - \frac{T_c}{T_h} = 1 - \frac{703 \text{ K}}{2143 \text{ K}} = 0.672 \text{ (or 67.2\%)} \qquad \Diamond$$

(b)   The actual efficiency of a heat engine is defined as $e = W_{eng}/Q_h$, where W_{eng} is the work done by the engine during some interval and Q_h is the energy absorbed from the hot reservoir during the same interval.

This engine absorbs $Q_h = 1.40 \times 10^5$ J from the hot reservoir each second and has an actual efficiency of $e = 0.420$ (42.0%). Therefore the work done by the engine each second (or the power delivered) is

$$\mathcal{P} = \frac{W_{eng}}{t} = \frac{eQ_h}{t} = \frac{(0.420)\left(1.40 \times 10^5 \text{ J}\right)}{1.00 \text{ s}} = 5.88 \times 10^4 \text{ W} = 58.8 \text{ kW} \qquad \Diamond$$

33. A nuclear power plant has an electrical power output of 1000 MW and operates with an efficiency of 33%. If excess energy is carried away from the plant by a river with a flow rate of 1.0×10^6 kg / s, what is the rise in temperature of the flowing water?

Solution   The efficiency of the plant is given by $e = W/Q_h$ where W is the net work output from the plant in some time interval Δt and Q_h is the energy input to the plant during the same interval. Thus, the required energy input is $Q_h = W/e$.

From conservation of energy, the excess energy that must be carried away by the cooling water of the river during this time interval is

$$Q_c = Q_h - W = \frac{W}{e} - W = \left(\frac{1-e}{e}\right)W$$

Since the work done during time Δt is given by $W = \mathcal{P}(\Delta t)$, where $\mathcal{P}$ is the power output, the excess energy becomes

$$Q_c = \left(\frac{1-e}{e}\right)W = \left(\frac{1-e}{e}\right)\mathcal{P}(\Delta t)$$

For a plant with a power output of 1000 MW and efficiency of 33%, the excess energy the river must carry away each second is

$$Q_c = \left(\frac{1-0.33}{0.33}\right)(1000 \times 10^6 \text{ J}/\text{s})(1.00 \text{ s}) = 2.0 \times 10^9 \text{ J}$$

The mass of cooling water flowing through the plant in this 1.00-s interval is $m = 1.0 \times 10^6$ kg. Therefore, the expected rise in temperature of the water is

$$\Delta T = \frac{Q_c}{mc_{\text{water}}} = \frac{2.03 \times 10^9 \text{ J}}{(1.0 \times 10^6 \text{ kg})(4186 \text{ J}/\text{kg}\cdot°\text{C})} = 0.49 °\text{C} \qquad \diamond$$

---

35. A freezer is used to freeze 1.0 L of water completely into ice. The water and the freezer remain at a constant temperature of $T = 0$ °C. Determine (a) the change in the entropy of the water and (b) the change in the entropy of the freezer.

Solution The change in the entropy of a thermodynamic system during a reversible, constant temperature process is given by $\Delta S = \Delta Q_r / T$ where ΔQ_r is the energy added to the system by heat and T is the **absolute** temperature of the system during this process.

As 1.0-L of liquid water at 0 °C is converted to ice, the quantity of energy transferred by heat **from** the water and **to** the freezer is

$$|\Delta Q_r| = mL_f = (\rho V)L_f = \left[(10^3 \text{ kg}/\text{m}^3)(1.0 \text{ L})\left(\frac{10^{-3} \text{ m}^3}{1 \text{ L}}\right)\right](3.33 \times 10^5 \text{ J}/\text{kg})$$

or $|\Delta Q_r| = 3.3 \times 10^5 \text{ J} = 330 \text{ kJ}$

The constant temperature of both the water and freezer during this process is $T = 0 °\text{C} = 273 \text{ K}$.

(a) Energy is transferred from the water during the freezing process, so $(\Delta Q_r)_{water} < 0$ and the change in the entropy of the water is

$$\Delta S_{water} = \frac{(\Delta Q_r)_{water}}{T} = \frac{-330 \text{ kJ}}{273 \text{ K}} = -1.2 \text{ kJ / K} \qquad \diamond$$

(b) Energy is transferred to the freezer by heat, so $(\Delta Q_r)_{freezer} > 0$

and $\quad \Delta S_{freezer} = \dfrac{(\Delta Q_r)_{freezer}}{T} = \dfrac{+330 \text{ kJ}}{273 \text{ K}} = +1.2 \text{ kJ / K} \qquad \diamond$

Note that in this **reversible** process, the total entropy of the isolated system consisting of (freezer plus water) is constant,

or $\quad \Delta S_{isolated \atop system} = \Delta S_{water} + \Delta S_{freezer} = 0$

---

41. Prepare a table like Table 12.4 for the following occurrence. You toss four coins into the air simultaneously. Record all the possible results of the toss in terms of the numbers of heads and tails that can result. (For example, HHTH and HTHH are two possible ways in which three heads and one tail can be achieved.) (a) On the basis of your table, what is the most probable result of a toss? (b) In terms of entropy, what is the most ordered state and (c) what is the most disordered?

Solution The desired table is shown below:

End Result	Possible Combinations	Total with same result
All Heads	HHHH	1
3 Heads, 1 Tail	THHH, HTHH, HHTH, HHHT	4
2 Heads, 2 Tails	TTHH, THTH, THHT, HTHT, HHTT, HTTH	6
1 Head, 3 Tails	HTTT, THTT, TTHT, TTTH	4
All Tails	TTTT	1

(a) The most probable result is seen to be 2 heads and 2 tails. Six combinations out of a total of 16 possible combinations will yield this result. The probability of obtaining 2 heads and 2 tails is thus 6/16. $\qquad \diamond$

(b) The results with the most order are those in which all coins are oriented the same way (either all heads or all tails). These two states are equally probable, each having a probability of 1/16. $\qquad \diamond$

(c) The result which allows the highest degree of disorder in the coin orientations is also the most probable state, (2 heads and 2 tails). $\qquad \diamond$

---

49. One object is at a temperature of T_h and another is at a lower temperature, T_c. Use the second law of thermodynamics to show that energy transfer by heat can only occur from the hotter to the colder object. Assume that the heat capacity of each object is large enough that its temperature is not measurably changed.

Solution The second law of thermodynamics states that the change in the total entropy of an isolated system can never be negative. In any thermodynamic process, this total entropy must either remain constant or increase.

Assume that, in an isolated system, energy of magnitude Q is transferred by heat from an object at absolute temperature T_c to a second object at absolute temperature $T_h > T_c$.

Then, the energy transfers for each object would be $Q_h = +Q$ and $Q_c = -Q$.

If the thermal capacity of each object is very large in comparison to Q, the temperatures of the objects remain essentially constant during this process. The entropy change associated with each object is then

$$\Delta S_h = \frac{Q_h}{T_h} = \frac{+Q}{T_h} \qquad \text{and} \qquad \Delta S_c = \frac{Q_c}{T_c} = \frac{-Q}{T_c}$$

The total change in entropy for this isolated system would be

$$\Delta S_{total} = \Delta S_h + \Delta S_c \qquad \text{or} \qquad \Delta S_{total} = \left(\frac{+Q}{T_h}\right) + \left(\frac{-Q}{T_c}\right) = Q\left(\frac{T_c - T_h}{T_h T_c}\right)$$

Since absolute temperatures are never negative, the product $T_h T_c$ is positive.

Since $T_h > T_c$, the factor $(T_c - T_h)/T_h T_c$ is negative.

Therefore, the only way this process can be in agreement with the second law $(\Delta S_{total} \geq 0)$ is for the factor Q to also be negative.

However, if $Q < 0$, then $Q = -|Q|$, and the energy transfers for each object are

$$Q_h = +Q = -|Q| < 0 \qquad \text{and} \qquad Q_c = -Q = -(-|Q|) = +|Q| > 0$$

That is, the second law of thermodynamics requires that the actual transfer of energy by heat be **from** the hotter body **to** the cooler body . ◊

55. One mole of neon gas is heated from 300 K to 420 K at constant pressure. Calculate (a) the energy Q transferred to the gas, (b) the change in the internal energy of the gas, and (c) the work done on the gas. Note that neon has a molar specific heat of $c = 20.79$ J / mol·K for a constant-pressure process.

Solution

(a) The energy transferred to this gas by heat is

$$Q = mc(\Delta T)$$

or $\quad Q = (1.00 \text{ mol})(20.79 \text{ J/mol·K})(420 \text{ K} - 300 \text{ K}) = 2.49 \times 10^3 \text{ J} = 2.49 \text{ kJ}$ ◊

(b) Neon is a noble gas, so it is monatomic, has negligible intermolecular forces, and closely approximates an ideal gas. The internal energy of the gas is thus

$$U = \frac{3}{2}PV = \frac{3}{2}nRT \qquad \text{(see Problem 12.1)}$$

The gas' internal energy changes by:

$$\Delta U = U_f - U_i = \frac{3}{2}nRT_f - \frac{3}{2}nRT_i = \frac{3}{2}nR(\Delta T)$$

or $\quad \Delta U = \frac{3}{2}(1.00 \text{ mol})(8.31 \text{ J/mol·K})(420 \text{ K} - 300 \text{ K}) = 1.50 \times 10^3 \text{ J} = 1.50 \text{ kJ}$ ◊

(c) The first law of thermodynamics, $\Delta U = Q + W$, gives the work done **on** the gas as

$$W = \Delta U - Q = 1.50 \times 10^3 \text{ J} - 2.49 \times 10^3 \text{ J} = -990 \text{ J}$$ ◊

Note that the gas expands when heated at constant pressure, so we should expect the work done on the gas to be negative.

57. Suppose a heat engine is connected to two energy reservoirs, one a pool of molten aluminum ($660 \, °C$) and the other a block of solid mercury ($-38.9 \, °C$). The engine runs by freezing 1.00 g of aluminum and melting 15.0 g of mercury during each cycle. The latent heat of fusion of aluminum is $3.97 \times 10^5 \, J/kg$, and that of mercury is $1.18 \times 10^4 \, J/kg$. (a) What is the efficiency of this engine? (b) How does the efficiency compare with that of a Carnot engine?

Solution

(a) During each cycle, the energy transferred to the engine from the molten aluminum is

$$Q_h = m_{Al}\left(L_f\right)_{Al} = \left(1.00 \times 10^{-3} \, kg\right)\left(3.97 \times 10^5 \, J/kg\right) = 397 \, J$$

Also, during each cycle, the energy exhausted from the engine to the frozen mercury is

$$Q_c = m_{Hg}\left(L_f\right)_{Hg} = \left(15.0 \times 10^{-3} \, kg\right)\left(1.18 \times 10^4 \, J/kg\right) = 177 \, J$$

Thus, the efficiency of the heat engine is given by

$$e = \frac{W_{eng}}{Q_h} = \frac{Q_h - Q_c}{Q_h} = 1 - \frac{Q_c}{Q_h} = 1 - \frac{177 \, J}{397 \, J} = 0.554 \quad \text{(or 55.4\%)} \qquad \Diamond$$

(b) The efficiency of a Carnot engine operating between reservoirs having temperatures of $T_h = 660 \, °C = 933 \, K$ and $T_c = -38.9 \, °C = 234 \, K$ is

$$e_C = \frac{T_h - T_c}{T_h} = 1 - \frac{T_c}{T_h} = 1 - \frac{234 \, K}{933 \, K} = 0.749 \quad \text{(or 74.9\%)} \qquad \Diamond$$

Chapter 13
VIBRATIONS AND WAVES

NOTES ON SELECTED CHAPTER SECTIONS

13.1   Hooke's Law

13.2   Elastic Potential Energy

Simple harmonic motion occurs when the net force along the direction of motion is a Hooke's law type of force; that is, when the net force is proportional to the displacement and in the opposite direction. It is necessary to define a few terms relative to harmonic motion:

1. **The amplitude, A, is the maximum distance that an object moves away from its equilibrium position.** In the absence of friction, an object will continue in simple harmonic motion and reach a maximum displacement equal to the amplitude on each side of the equilibrium position during each cycle.

2. **The period, T, is the time it takes the object to execute one complete cycle of the motion.**

3. **The frequency, f, is the number of cycles or vibrations per unit of time.**

Oscillatory motions are exhibited by many physical systems such as a mass attached to a spring, a pendulum, atoms in a solid, stringed musical instruments, and electrical circuits driven by a source of alternating current. **Simple harmonic motion** of a mechanical system corresponds to the oscillation of an object between two points for an indefinite period of time, with no loss in mechanical energy.

An object exhibits simple harmonic motion if the net external force acting on it is a linear restoring force.

13.5   Position, Velocity, and Acceleration as a Function of Time

The position (x), velocity (v), and acceleration (a) of an object moving with simple harmonic motion are shown in the three graphs below. In this particular case, the object was released from rest when it was a maximum distance (amplitude) from the equilibrium position.

$$x = A\cos(2\pi f t)$$

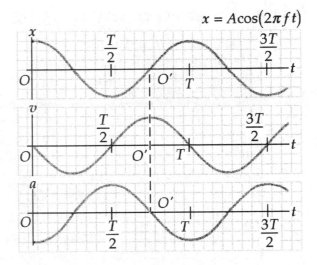

Shown here, from top to bottom, are graphs of displacement, velocity, and acceleration versus time for an object moving with simple harmonic motion under the initial conditions that $x_0 = A$ and $v_0 = 0$ at $t = 0$.

The most common system which undergoes simple harmonic motion is the mass-spring system shown in the figure at the right. The mass is assumed to move on a horizontal, frictionless surface. The point $x = 0$ is the equilibrium position of the mass; that is, the point where the mass would reside if left undisturbed. In this position, there is no horizontal force on the mass. When the mass is displaced a distance x from its equilibrium position, the spring produces a linear restoring force given by Hooke's law, $F = -kx$, where k is the force constant of the spring, and has SI units of N/m. The minus sign means that **F** is to the left when the displacement x is positive, whereas **F** is to the right when x is negative. In other words, **the direction of the force F is always toward the equilibrium position**.

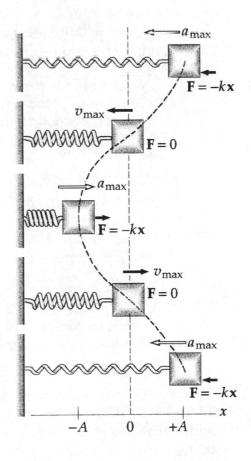

13.6 Motion of a Pendulum

A **simple pendulum** consists of a mass m attached to a light string of length L as shown in the figure. When the angular displacement θ is small during the entire motion (less than about 15°), the pendulum exhibits simple harmonic motion. In this case, the resultant force acting on the mass m equals the component of weight **tangent** to the circular path followed by the mass. The magnitude of the resultant force equals $mg\sin\theta$. Since this force is always directed toward $\theta = 0$, it corresponds to a restoring force.

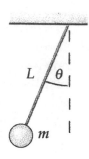

The period depends only on the length of the pendulum and the acceleration of gravity. The period does not depend on mass, so we conclude that all simple pendula of equal length oscillate with the same frequency or period.

13.7 Damped Oscillations

Damped oscillations occur in realistic systems in which retarding forces such as friction are present. These forces will reduce the amplitudes of the oscillations with time, since mechanical energy is continually transferred from the oscillating system. Depending on the value of the frictional (retarding) force three distinct types of damping can be identified:

Underdamped: In this case, the retarding force is small compared to the restoring force. Oscillations continue with the frequency of the undamped system (passing through the equilibrium point twice during each oscillation). The amplitude of the motion decreases exponentially with time until the oscillations cease at zero amplitude.

Critically damped: With an increase in the retarding force the system does not oscillate and returns to the equilibrium position in the shortest possible time without passing through the equilibrium point.

Overdamped: This mode is similar to critical damping and occurs with further increase in the frictional force. In this case the system returns to equilibrium without passing through the equilibrium point but requires a longer time to do so.

It is possible to compensate for the energy lost in a damped oscillator by adding an additional driving force that does positive work on the system. This additional energy supplied to the system must equal the energy lost due to friction in order to maintain constant amplitude.

13.8   Wave Motion

The production of mechanical waves require: (1) an **elastic medium** which can be disturbed, (2) an **energy source** to provide a disturbance or deformation in the medium, and (3) a physical mechanism by way of which adjacent portions of the medium can **influence** each other. The three parameters important in characterizing waves are (1) wavelength, (2) frequency, and (3) wave velocity.

13.9   Types of Waves

Transverse waves are those in which particles of the disturbed medium move along a direction which is perpendicular to the direction of the wave velocity. For **longitudinal waves**, the particles of the medium undergo displacements which are parallel to the direction of wave motion.

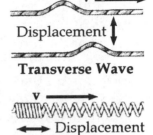

Transverse Wave

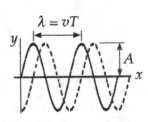

Longitudinal Wave

13.10   Frequency, Amplitude, and Wavelength

Consider a wave traveling in the x direction on a very long string. Each particle along the string oscillates in simple harmonic motion along the y direction with a frequency equal to that of the source producing the vibration. The maximum distance the string is displaced above or below the equilibrium value is called the **amplitude**, A, of the wave. The distance between two successive points along the string that behave identically is called the **wavelength**, λ. The wave will advance along the string a distance of one wavelength in a time interval equal to one period of vibration, T. This is also the time required for any point in the medium to complete one cycle of its vibration.

13.11   The Speed of Waves on Strings

For linear waves, the **velocity** of **mechanical waves** depends only on the physical properties of the medium through which the disturbance travels. In the case of **waves on a string**, the velocity depends on the tension in the string and the mass per unit length (linear mass density).

13.12   Interference of Waves

If two or more waves are moving through a medium, the **resultant wave function is the algebraic sum of the wave functions of the individual waves**. Two traveling waves can pass through each other without being destroyed or altered.

13.13   Reflection of Waves

Whenever a traveling wave reaches a boundary, part or all of the wave is reflected. If the wave is traveling along a string and is reflected from a "fixed" end, the reflected pulse is inverted. By contrast, a pulse is reflected without inversion at the "free" end of a string.

EQUATIONS AND CONCEPTS

The force exerted by a spring on a mass attached to the spring and displaced a distance x from the unstretched position is given by Hooke's law. The force constant, k, is always positive and has a value which corresponds to the relative stiffness of the spring. The negative sign means that the force exerted on the mass is always directed opposite the displacement; the force is a restoring force, always directed toward the equilibrium position.

$$F_s = -kx \qquad (13.1)$$

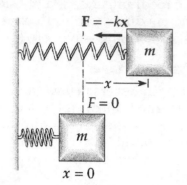

An object exhibits simple harmonic motion when the net force along the direction of motion is proportional to the displacement and oppositely directed.

Comment on simple harmonic motion.

This equation gives the acceleration of an object in simple harmonic motion as a function of position. Note that when the oscillating mass is at the equilibrium position ($x = 0$), the acceleration $a = 0$. The acceleration has its maximum magnitude when the displacement of the mass is maximum, $x = \pm A$ (amplitude).

$$a = -\left(\frac{k}{m}\right)x \qquad (13.2)$$

Work must be done by an external applied force in order to stretch or compress a spring. This work results in energy, called elastic potential energy, being stored in the spring.

$$PE_s \equiv \frac{1}{2}kx^2 \qquad (13.3)$$

The spring force is conservative; hence, in the absence of friction or other nonconservative forces, the total mechanical energy (kinetic, gravitational potential, and elastic potential) of the spring-mass system remains constant.

$$(KE + PE_g + PE_s)_i =$$
$$(KE + PE_g + PE_s)_f \qquad (13.4)$$

The speed of an object in simple harmonic motion is a maximum at $x = 0$; the speed is zero when the mass is at the points of maximum displacement ($x = \pm A$).

$$v = \pm \sqrt{\frac{k}{m}(A^2 - x^2)} \qquad (13.6)$$

The period of an object in simple harmonic motion is the time required to complete a full cycle of its motion.

$$T = 2\pi \sqrt{\frac{m}{k}} \qquad (13.8)$$

The frequency, the number of cycles per unit time, is the reciprocal of the period. The units of frequency are hertz (Hz).

$$f = \frac{1}{T} \qquad (13.9)$$

$$f = \frac{1}{2\pi}\sqrt{\frac{k}{m}} \qquad (13.10)$$

These equations represent the position of an object moving along the x-axis in simple harmonic motion as a function of time.

$$x = A\cos(\omega t) \qquad (13.12)$$

$$x = A\cos(2\pi f t) \qquad (13.14)$$

$\text{Cos}(0) = 1$. Hence in Equations 13.12 and 13.14, $x = A$ when $t = 0$. Therefore, the particular form of the position equations shown above assumes that the vibrating object is at the point of maximum displacement when $t = 0$.

Comment on
Equation 13.12 and Equation 13.14.

$$t = 0 \;\Big]\; \text{Initial}$$
$$x = 0 \qquad v_0 = 0 \;\Big]\; \text{Conditions}$$

ω is the angular frequency (rad/s) of an object in simple harmonic motion and f is the number of oscillations completed per unit time measured in hertz (Hz).

$$\omega = 2\pi f \qquad (13.13)$$

The period of oscillation of a simple pendulum depends only on its length, L, and the acceleration due to gravity, g. The period does not depend on the mass; and to a good approximation, the period does not depend on the amplitude, θ_{max}, within the range of small amplitudes.

$$T = 2\pi\sqrt{\frac{L}{g}}$$

(13.15)

The wave speed, v, is the rate at which the disturbance or pulse moves along the direction of travel of the wave.

Wave speed, v

$$v = f\lambda$$

(13.16)

The wave speed in a stretched string depends on the tension in the string and the linear density (mass per unit length). For any mechanical wave, the speed depends only on the properties of the medium through which the wave travels.

$$v = \sqrt{\frac{F}{\mu}}$$

(13.17)

REVIEW CHECKLIST

▷ Describe the general characteristics of a system in simple harmonic motion; and define amplitude, period, frequency, and displacement.

▷ Define the following terms relating to wave motion: frequency, wavelength, velocity, and amplitude; and express a given harmonic wave function in several alternative forms involving different combinations of the wave parameters: wavelength, period, phase velocity, angular frequency, and harmonic frequency.

▷ Given a specific wave function for a harmonic wave, obtain values for the characteristic wave parameters: A, ω, and f.

▷ Make calculations which involve the relationships between wave speed and the inertial and elastic characteristics of a string through which the disturbance is propagating.

▷ Define and describe the following wave associated phenomena: superposition, phase, interference, and reflection.

SOLUTIONS TO SELECTED END-OF-CHAPTER PROBLEMS

3. A ball dropped from a height of 4.00 m makes a perfectly elastic collision with the ground. Assuming no mechanical energy is lost due to air resistance, (a) show that the motion is periodic and (b) determine the period of the motion. (c) Is the motion simple harmonic? Explain.

Solution

(a) Since the collision is perfectly elastic ($KE_{after} = KE_{before}$) and the recoil of the Earth may be neglected, the speed of the ball as it leaves the ground after a collision is the same as its speed immediately before the collision. The ball rebounds to the height from which it was initially dropped, comes to rest, and then falls again.

The ball repeats this motion continuously, and thus undergoes a periodic motion.◊

(b) The period (time for one complete cycle) of this motion is the total elapsed time from when the ball starts at rest 4.00 m above ground until it returns to rest at 4.00 m high following a bounce.

For the downward motion, $v_i = 0$, $\Delta y = -4.00$ m, $a_y = -g = -9.80$ m / s^2

Thus,
$$\Delta y = v_i t + \frac{1}{2} a_y t^2$$

gives the time for the ball to reach the ground: $\quad t = \sqrt{\dfrac{2(-4.00 \text{ m})}{-9.80 \text{ m / s}^2}} = 0.904$ s

The speed of the ball just before impact is

$$v_1 = v_i + a_y t_1 = 0 + \left(-9.80 \text{ m / s}^2\right)(0.904 \text{ s}) = -8.85 \text{ m / s}$$

On the upward flight, $\quad v_2 = -v_1 = +8.85$ m / s, $\quad v_f = 0$, and $\quad \Delta y = +4.00$ m

The average velocity for this part of the motion is $\quad \bar{v} = \frac{1}{2}\left(v_2 + v_f\right) = 4.43$ m / s

The time to return to the initial height is $\quad t_2 = \dfrac{\Delta y}{\bar{v}} = \dfrac{4.00 \text{ m}}{4.43 \text{ m / s}} = 0.904$ s

The total time for this complete cycle of the motion is then $T = t_1 + t_2 = 1.81$ s $\quad$ ◊

(c) Simple harmonic motion is motion produced by a Hooke's law type force. Such a force is proportional to the displacement from an equilibrium position and is in the direction opposite to the displacement. This ball has a constant downward force, $w = mg$, acting on it while it is in the air. When the ball is at ground level, an upward impulsive force acts on it. These forces are not Hooke's law type forces. Thus, the motion **is not** simple harmonic. $\quad$ ◊

9. A child's toy consists of a piece of plastic attached to a spring (Fig. P13.9). The spring is compressed against the floor a distance of 2.00 cm, and the toy is released. If the toy has a mass of 100 g and rises to a maximum height of 60.0 cm, estimate the force constant of the spring.

Solution When the toy is at rest against the floor with the spring compressed 2.00 cm, the total mechanical energy of the toy is

Figure P13.9

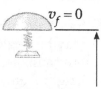

$$E_i = KE_i + PE_{g,i} + PE_{s,i} = \frac{1}{2}mv_i^2 + mgy_i + \frac{1}{2}kx_i^2$$

or $\quad E_i = 0 + 0 + \frac{1}{2}kx_i^2 = \frac{1}{2}kx_i^2$

where the zero gravitational potential energy level has been chosen at the floor. When the toy comes to rest temporarily at its maximum height, the spring is no longer compressed and the total total mechanical energy is

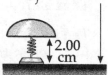

$$E_f = KE_f + PE_{g,f} + PE_{s,f} = \frac{1}{2}mv_f^2 + mgy_f + \frac{1}{2}kx_f^2 = 0 + mgy_f + 0 = mgy_f$$

The forces acting on the toy during the intervening time interval are a gravitational force and a spring force, both conservative forces.

Therefore, the mechanical energy is constant, or $E_i = E_f$.

This gives $\frac{1}{2}kx_i^2 = mgy_f$, so the force constant is

$$k = \frac{2mgy_f}{x_i^2} = \frac{2(0.100 \text{ kg})(9.80 \text{ m / s}^2)(0.600 \text{ m})}{(2.00 \times 10^{-2} \text{ m})^2} = 2.94 \times 10^3 \text{ N / m} \qquad \Diamond$$

15. A 0.40‑kg object connected to a light spring with a spring constant of 19.6 N / m oscillates on a frictionless horizontal surface. If the spring is compressed 4.0 cm and released from rest, determine (a) the maximum speed of the object, (b) the speed of the object when the spring is compressed 1.5 cm, and (c) the speed of the object when the spring is stretched 1.5 cm. (d) For what value of x does the speed equal one-half the maximum speed?

Solution

Choosing the horizontal surface as the reference level $(y = 0)$, $PE_g = mgy = 0$ at all times and the total energy at any point in the motion is

$$E = KE + PE_s = \frac{1}{2}mv^2 + \frac{1}{2}kx^2$$

The only force doing work on the object is a conservative spring force. Therefore, the total energy is constant, or $E = E_i$ where E_i is the energy when first released.

Since the object starts from rest, $\qquad\qquad E_i = 0 + \frac{1}{2}kx_i{}^2 = \frac{1}{2}kx_i{}^2$

and the speed of the object when it is at a displacement x from the equilibrium position is found from

$$\frac{1}{2}mv^2 + \frac{1}{2}kx^2 = \frac{1}{2}kx_i{}^2 \qquad \text{or} \qquad v = \sqrt{\left(x_i^2 - x^2\right)k/m} \qquad\qquad \text{[1]}$$

(a) Observing Equation [1], it is clear that the object has maximum speed when $x = 0$ (i.e., at the equilibrium position).

$$\text{Thus,} \quad v_{max} = \sqrt{\frac{\left(x_i{}^2 - 0\right)k}{m}} = \sqrt{\frac{(0.040 \text{ m})^2(19.6 \text{ N / m})}{0.40 \text{ kg}}} = 0.28 \text{ m / s} = 28 \text{ cm / s} \qquad \Diamond$$

(b) When the spring is compressed 1.5 cm, $x = -0.015$ m and Equation [1] gives the speed as

$$v = \sqrt{\left[(0.040 \text{ m})^2 - (-0.015 \text{ m})^2\right]\frac{19.6 \text{ N / m}}{0.40 \text{ kg}}} = 0.26 \text{ m / s} = 26 \text{ cm / s} \qquad \Diamond$$

(c) When the spring is stretched 1.5 cm, then $x = +0.015$ m and

$$v = \sqrt{\left[(0.040 \text{ m})^2 - (+0.015 \text{ m})^2\right]\frac{19.6 \text{ N / m}}{0.40 \text{ kg}}} = 0.26 \text{ m / s} = 26 \text{ cm / s} \qquad \Diamond$$

(d) When $v = \frac{1}{2}v_{max}$, Equation [1] gives $\qquad \sqrt{\left(x_i{}^2 - x^2\right)k/m} = \frac{1}{2}\sqrt{x_i{}^2 k/m}$

or $\qquad\qquad\qquad\qquad\qquad\qquad\qquad 4\left(x_i{}^2 - x^2\right) = x_i{}^2$

which reduces to $\qquad\qquad\qquad\qquad x = \sqrt{\frac{3x_i{}^2}{4}} = \frac{\sqrt{3}(4.0 \text{ cm})}{2} = 3.5 \text{ cm} \qquad \Diamond$

23. A spring stretches 3.9 cm when a 10-g object is hung from it. The object is replaced with a block of mass 25 g which oscillates in simple harrmonic motion. Calculate the period of motion.

Solution

A simple harmonic oscillator moves under the influence of a Hooke's law force, $F_s = -kx$ where x is the displacement from equilibrium.

The spring constant is given by
$$k = |F_s|/|x| = F_s/x$$

When a 10-g object hangs from the spring, the force stretching the spring is

$$F_s = mg = (10 \times 10^{-3} \text{ kg})(9.80 \text{ m/s}^2) = 0.098 \text{ N}$$

If this produces an elongation of
$$x = 3.9 \text{ cm} = 0.039 \text{ m}$$

the spring constant is
$$k = \frac{F_s}{x} = \frac{0.098 \text{ N}}{0.039 \text{ m}} = 2.5 \text{ N/m}$$

When a total mass of 25 g is attached to this spring and set into oscillation, the period of the resulting simple harmonic motion will be

$$T = 2\pi\sqrt{\frac{m}{k}} = 2\pi\sqrt{\frac{25 \times 10^{-3} \text{ kg}}{2.5 \text{ N/m}}} = 0.63 \text{ s} \qquad \Diamond$$

---

29. Given that $x = A\cos(\omega t)$ is a sinusoidal function of time, show that v (velocity) and a (acceleration) are also sinusoidal functions of time. (**Hint:** Use Equations 13.6 and 13.2.)

Solution The total energy of a simple harmonic oscillator is

$$E = KE + PE_s = \frac{1}{2}mv^2 + \frac{1}{2}kx^2$$

Since the force acting on the oscillating mass is a **conservative** force $(F_s = -kx)$, the total energy is constant. At the maximum displacement from equilibrium $(x = A)$, the speed is $v = 0$, so the total energy is

$$E = 0 + \frac{1}{2}kA^2 = \frac{1}{2}kA^2$$

Thus, at arbitrary displacement x, we have $\qquad \frac{1}{2}mv^2 + \frac{1}{2}kx^2 = \frac{1}{2}kA^2$

which gives the velocity as $\qquad v = \pm\sqrt{\frac{k}{m}\left(A^2 - x^2\right)} = \pm\sqrt{\omega^2\left(A^2 - x^2\right)} = \pm\omega\sqrt{A^2 - x^2}$

From Newton's second law, the acceleration is $\qquad \mathbf{a} = \mathbf{F}/m$

Thus, when $\mathbf{F} = \mathbf{F}_s = -k\mathbf{x}$, the acceleration is given by $\qquad a = -\frac{k}{m}x = -\omega^2 x$

For a simple harmonic oscillator,
the position is a sinusoidal function of time, $\qquad x = A\cos(\omega t)$

The velocity is then $\qquad v = \pm\omega\sqrt{A^2 - x^2} = \pm\omega A\sqrt{1 - \cos^2(\omega t)} = \pm\omega A\sin(\omega t)$

which is another sinusoidal function of time. $\qquad\qquad\qquad\qquad\qquad\qquad\qquad \lozenge$

The acceleration of the simple harmonic oscillator is given by

$$a = -\omega^2 x = -\left(\omega^2 A\right)\cos(\omega t), \text{ also a sinusoidal function of time.} \qquad \lozenge$$

---

35. The free-fall acceleration on Mars is $3.7 \text{ m}/\text{s}^2$. (a) What length pendulum has a period of 1 s on Earth? What length pendulum would have a 1-s period on Mars? (b) An object is suspended from a spring with spring constant $10 \text{ N}/\text{m}$. Find the mass suspended from this spring that would result in a period of 1 s on Earth and on Mars.

Solution

(a) The period of a simple pendulum is $T = 2\pi\sqrt{L/g}$ where L is the length of the pendulum and g is the acceleration due to gravity at the pendulum's location.

Thus, if a pendulum has a period of $T = 1.0$ s on Earth where $g = 9.8 \text{ m}/\text{s}^2$, its length is

$$L_{\text{Earth}} = \frac{g_{\text{Earth}}T^2}{4\pi^2} = \frac{\left(9.8 \text{ m/s}^2\right)\left(1.0 \text{ s}\right)^2}{4\pi^2} = 0.25 \text{ m} = 25 \text{ cm} \qquad \lozenge$$

On Mars, where $g = 3.7 \text{ m}/\text{s}^2$, the length of a pendulum with a 1.0-s period is

$$L_{\text{Mars}} = \frac{g_{\text{Mars}}T^2}{4\pi^2} = \frac{\left(3.7 \text{ m/s}^2\right)\left(1.0 \text{ s}\right)^2}{4\pi^2} = 0.094 \text{ m} = 9.4 \text{ cm} \qquad \lozenge$$

(b) The period of vibration for an object suspended from a spring is $T = 2\pi\sqrt{m/k}$ where k is the force constant of the spring and m is the mass of the object. Thus, to have a period of $T = 1.0$ s when suspended from a spring with a force constant $k = 10$ N / m, the required mass is

$$m = \frac{kT^2}{4\pi^2} = \frac{(10 \text{ N/m})(1.0 \text{ s})^2}{4\pi^2} = 0.25 \text{ kg}$$

Both k and m are constants of the system and do not depend on the location of the system. Therefore, the same mass is needed on Earth and Mars, or

$$m_{\text{Earth}} = m_{\text{Mars}} = 0.25 \text{ kg} \qquad \Diamond$$

Note that while the same mass is needed on Earth and on Mars, the weight $w = mg$ will be different on the two planets. Thus, the oscillations occur about a different equilibrium position on Mars than on Earth.

---

41. A harmonic wave is traveling along a rope. It is observed that the oscillator that generates the wave completes 40.0 vibrations in 30.0 s. Also, a given maximum travels 425 cm along the rope in 10.0 s. What is the wavelength?

Solution The frequency of the oscillator, and the frequency that wave-crests (maxima) start down the rope, is

$$f = \frac{40.0 \text{ vibrations}}{30.0 \text{ s}} = \frac{4}{3} \text{ Hz}$$

The speed of the wave in the rope is $\qquad v = \dfrac{425 \text{ cm}}{10.0 \text{ s}} = 42.5 \text{ cm/s}$

The wave's wavelength, frequency, and wave speed are related by $v = \lambda f$.

Therefore, the wavelength is $\qquad \lambda = \dfrac{v}{f} = \dfrac{42.5 \text{ cm/s}}{4/3 \text{ s}^{-1}} = 31.9 \text{ cm} \qquad \Diamond$

---

47. A simple pendulum consists of a ball of mass 5.00 kg hanging from a uniform string of mass 0.0600 kg and length L. If the period of oscillation for the pendulum is 2.00 s, determine the speed of a transverse wave in the string when the pendulum hangs vertically.

Solution The period of a simple pendulum is $T = 2\pi\sqrt{L/g}$. Thus, if this pendulum has a period of $T = 2.00$ s, its length must be

$$L = \frac{gT^2}{4\pi^2} = \frac{\left(9.80 \text{ m/s}^2\right)\left(2.00 \text{ s}\right)^2}{4\pi^2} = 0.993 \text{ m}$$

The mass per unit length of the string is then $\mu = \dfrac{m_{\text{string}}}{L} = \dfrac{0.0600 \text{ kg}}{0.993 \text{ m}} = 0.0604 \text{ kg} / \text{m}$

When the pendulum hangs vertical and stationary, the tension in the string is

$$F = m_{\text{ball}}g = \left(5.00 \text{ kg}\right)\left(9.80 \text{ m} / \text{s}^2\right) = 49.0 \text{ N}$$

The speed of a transverse wave in the string under these conditions will be

$$v = \sqrt{\frac{F}{\mu}} = \sqrt{\frac{49.0 \text{ N}}{0.0604 \text{ kg} / \text{m}}} = 28.5 \text{ m} / \text{s} \qquad \diamond$$

---

53. A wave of amplitude 0.30 m interferes with a second wave of amplitude 0.20 m traveling in the same direction. What are the (a) largest and (b) smallest resultant amplitudes that can occur, and under what conditions will these maxima and minima occur?

Solution When two waves meet at a point, the resultant displacement at that point for that instant in time is found by adding the displacements of the individual waves at that point and time.

(a) The largest magnitude displacement at the point will occur if the two waves each have maximum displacements (equal to the amplitudes of the waves) **in the same direction** at the instant they meet. That is, it will occur when the two waves meet in phase and interfere constructively.   In this case the magnitude of the displacement at the point is

$$\left|\text{resultant displacment}\right| = A_1 + A_2 = 0.30 \text{ m} + 0.20 \text{ m} = 0.50 \text{ m} \qquad \diamond$$

(b) The smallest magnitude displacement occurs at the point if the two waves each have maximum displacements **in opposite directions** at the instant they meet. That is, when the two waves meet out of phase and interfere destructively. In that case the magnitude of the displacement at the point is

$$\left|\text{resultant displacment}\right| = A_1 - A_2 = 0.30 \text{ m} - 0.20 \text{ m} = 0.10 \text{ m} \qquad \diamond$$

---

57. A 3.00-kg object is fastened to a light spring with the intevening cord passing over a pulley (Fig. P13.57). The pulley is frictionless, and its inertia may be neglected. The object is released from rest when the spring is unstretched. If the object drops 10.0 cm before stopping, find (a) the spring constant of the spring and (b) the speed of the object when it is 5.00 cm below its starting point.

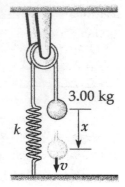

Solution Choose the zero gravitational potential energy level where the object starts from rest and the spring is unstretched. The pulley is frictionless with negligible inertia, and conservation of energy gives the energy after the spring is stretched a distance x as $E = E_i$ where E_i is the energy at $x = 0$.

Figure P13.57

Thus, $KE + PE_g + PE_s = KE_i + PE_{g,i} + PE_{s,i}$

or $\frac{1}{2}mv^2 - mgx + \frac{1}{2}kx^2 = 0 + 0 + 0$

and the speed of the object after it has dropped a distance x is

$$v = \sqrt{x\left(2g - \frac{kx}{m}\right)} = \sqrt{x\left(19.6 \text{ m / s}^2 - \frac{kx}{m}\right)} \qquad [1]$$

(a)   Since the object comes to rest $(v = 0)$ when $x = x_{max} = 10.0 \text{ cm} = 0.100 \text{ m}$, the energy equation $(E = \frac{1}{2}mv^2 - mgx + \frac{1}{2}kx^2 = 0)$ gives the force constant as

$$k = \frac{2mg}{x_{max}} = \frac{2(3.00 \text{ kg})(9.80 \text{ m / s}^2)}{0.100 \text{ m}} = 588 \text{ N / m} \qquad \Diamond$$

(b)   When $x = 5.00 \text{ cm} = 0.0500 \text{ m}$, Equation [1] gives the speed of the object as

$$v = \sqrt{(0.0500 \text{ m})\left[19.6 \text{ m / s}^2 - \frac{(588 \text{ N / m})(0.0500 \text{ m})}{3.00 \text{ kg}}\right]} = 0.700 \text{ m / s} \qquad \Diamond$$

Note that this is also the maximum speed of the object since $x = 5.00$ cm is the equilibrium position of the object.

---

61. A 2.00-kg block hangs without vibrating at the end of a spring ($k = 500 \text{ N / m}$) that is attached to the ceiling of an elevator car. The car is rising with an upward acceleration of $g / 3$ when the acceleration suddenly ceases (at $t = 0$). (a) What is the angular frequency of oscillation of the block after the acceleration ceases? (b) By what amount is the spring stretched during the time that the elevator car is accelerating? This distance will be the amplitude of the ensuing oscillation of the block.

Solution

(a) The angular frequency of oscillation of an object attached to the end of a spring is given by $\omega = \sqrt{k/m}$ where k is the force constant of the spring and m is the mass of the object. Thus, for this system, the angular frequency is

$$\omega = \sqrt{\frac{500 \text{ N/m}}{2.00 \text{ kg}}} = \sqrt{250 \text{ s}^{-2}} = 15.8 \text{ rad/s} \qquad \Diamond$$

(b) The sketch at the right gives a free-body diagram of the block while the elevator (as well as its contents, including the block) is accelerating upward at $a_y = +g/3$. Applying Newton's second law to the block gives $\Sigma F_y = F_s - mg = m(+g/3)$, so the tension in the spring must be

$$F_s = \frac{4}{3}mg = \frac{4}{3}(2.00 \text{ kg})(9.80 \text{ m/s}^2) = 26.1 \text{ N}$$

From Hooke's law, the amount the spring is stretched is then

$$x = \frac{F_s}{k} = \frac{26.1 \text{ N}}{500 \text{ N/m}} = 5.23 \times 10^{-2} \text{ m} = 5.23 \text{ cm} \qquad \Diamond$$

---

67. An object of mass $m_1 = 9.0$ kg is in equilibrium while connected to a light spring of constant $k = 100$ N/m that is fastened to a wall as in Figure P13.67a. A second object of mass $m_2 = 7.0$ kg is slowly pushed up against m_1, compressing the spring by the amount $A = 0.20$ m, as shown in Figure P13.67b. The system is then released, causing both objects to start moving to the right on the frictionless surface. (a) When m_1 reaches the equilibrium point, m_2 loses contact with m_1 (Fig. P13.67c) and moves to the right with speed v. Determine the value of v. (b) How far apart are the objects when the spring is fully stretched for the first time (Fig. P13.67d)? (**Hint:** First determine the period of oscillation and the amplitude of the m_1-spring system after m_2 loses contact with m_1.)

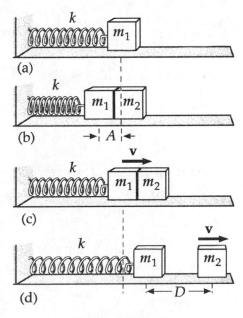

Figure P13.67

Solution

(a) Using conservation of energy from the moment of release to the instant of separation gives

$$\left(KE + PE_g + PE_s\right)_f = \left(KE + PE_g + PE_s\right)_i$$

or $\frac{1}{2}(m_1 + m_2)v_{max}^2 + 0 + 0 = 0 + 0 + \frac{1}{2}kA^2$

Thus, $v_{max} = A\sqrt{\dfrac{k}{m_1 + m_2}} = (0.20 \text{ m})\sqrt{\dfrac{100 \text{ N / m}}{(9.0 + 7.0) \text{ kg}}} = 0.50 \text{ m / s}$ ◊

(b) After the two blocks separate, m_1 oscillates with amplitude A' found by applying

$$\left(KE + PE_g + PE_s\right)_f = \left(KE + PE_g + PE_s\right)_i$$

to the $(m_1 + \text{spring})$ system from the moment of separation until the spring is fully stretched the first time.

$$0 + 0 + \frac{1}{2}kA'^2 = \frac{1}{2}m_1 v_{max}^2 + 0 + 0$$

$$A' = v_{max}\sqrt{\frac{m_1}{k}} = (0.50 \text{ m / s})\sqrt{\frac{9.0 \text{ kg}}{100 \text{ N / m}}} = 0.15 \text{ m}$$

The period of this oscillation is

$$T = 2\pi\sqrt{\frac{m_1}{k}} = 2\pi\sqrt{\frac{9.0 \text{ kg}}{100 \text{ N / m}}} = 1.9 \text{ s}$$

so the spring is fully stretched for the first time at $t = \frac{1}{4}T = 0.47$ s after separation. During this time, m_2 has moved distance $x = v_{max}t$ from the point of separation.

Thus, the distance separating the two blocks at this instant is

$$D = v_{max}t - A' = (0.50 \text{ m / s})(0.47 \text{ s}) - 0.15 \text{ m} = 0.086 \text{ m} = 8.6 \text{ cm}$$ ◊

Chapter 14

SOUND

NOTES ON SELECTED CHAPTER SECTIONS

14.1 Producing a Sound Wave

14.2 Characteristics of Sound Waves

Sound waves, which have as their source vibrating objects, are **longitudinal waves** traveling through a medium such as air. The particles of the medium **oscillate back and forth along the direction in which the wave travels**. This is in contrast to a transverse wave, in which the vibrations of the medium are at right angles to the direction of travel of the wave.

A sound wave traveling through air creates alternating regions of high and low molecular density and air pressure. A region of high density and air pressure is called a **compression** or **condensation**; a region of lower-than-normal density is referred to as a **rarefaction**. A sinusoidal curve can be used to represent a sound wave. There are crests in the sinusoidal wave at the points where the sound wave has condensations, and troughs where the sound wave has rarefactions.

14.4 Energy and Intensity of Sound Waves

The **intensity** of a wave is the rate at which sound energy flows through a unit area perpendicular to the direction of travel of the wave. The faintest sounds the human ear can detect have an intensity of about $1 \times 10^{-12} \, \text{W} / \text{m}^2$. This intensity is called the **threshold of hearing**. The loudest sounds the ear can tolerate, at the **threshold of pain**, have an intensity of about $1 \, \text{W} / \text{m}^2$. The sensation of loudness is approximately logarithmic in the human ear, and the relative intensity of a sound is called the **intensity level** or **decibel level**.

14.5 Spherical and Plane Waves

The intensity of a **spherical wave produced by a point source** is proportional to the average power emitted and inversely proportional to the square of the distance from the source.

14.6 The Doppler Effect

In general, the Doppler effect is experienced whenever there is relative motion between source and observer. When the source and observer are moving toward each other, the frequency heard by the observer is higher than the frequency of the source. When the source and observer are moving away from each other, the observer hears a frequency lower than the source frequency.

14.8 Standing Waves

Standing waves can be set up in a string by the superposition of two wave trains traveling in opposite directions along the string. This can occur when waves reflected from one end of the string interfere with the incident wave train, under the condition that the wavelengths (and therefore frequencies) are properly matched to the length of the string. The string has a number of natural patterns of vibration, called **normal modes**. Each normal mode has a **characteristic frequency**. The lowest of these frequencies is called the **fundamental frequency**, which together with the higher frequencies form a **harmonic series**.

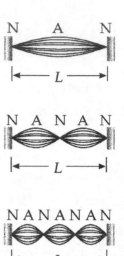

The figure on the right is a schematic representation of standing waves in a string of length L. In each case the envelope represents successive positions of the string during one complete cycle. Imagine that you observe the string vibrate while it is illuminated with a strobe light. The first three normal modes are shown. The points of zero displacement are called **nodes**; the points of maximum displacement are called **antinodes**.

14.9 Forced Vibrations and Resonance

Consider a system (for example a mass-spring system) which has a natural frequency of vibration, f_0, and is driven or pushed back and forth with a periodic force whose frequency is f. This type of motion is referred to as a **forced vibration**. Its amplitude reaches a maximum when the frequency of the driving force equals the natural frequency of the system, f_0, called the **resonant frequency** of the system. Under this condition, the system is said to be in **resonance**.

14.10 Standing Waves in Air Columns

Sound sources can be used to produce **longitudinal** standing waves in air columns. The phase relationship between incident and reflected waves depends on whether or not the reflecting end of the air column is open or closed. This gives rise to two sets of possible standing wave conditions:

In a pipe **open at both ends**, the natural frequencies of vibration form a series in which **all harmonics are present** and are equal to integral multiples of the fundamental.

In a pipe **closed at one end and open at the other**, only odd **harmonics are present**

14.11 Beats

Consider a type of interference effect that results from the superposition of **two waves with slightly different frequencies**. In this situation, at some fixed point the waves are periodically in and out of phase, corresponding to an alternation in time between constructive and destructive interference. A listener hears an alternation in loudness, known as **beats**. The number of beats per second, or **beat frequency, equals the difference in frequency between the two sources**.

EQUATIONS AND CONCEPTS

A sound wave propagates through a gas or liquid as a compressional wave. The speed of the sound wave depends on the value of the bulk modulus, B (an elastic property), and the equilibrium density, ρ (an inertial property), of the material through which it is traveling.

$$v = \sqrt{\frac{B}{\rho}} \tag{14.1}$$

$$B = -\frac{\Delta P}{\Delta V/V} \tag{14.2}$$

The speed of sound (or any longitudinal wave) in a solid depends on the value of Young's modulus and the density of the material.

$$v = \sqrt{\frac{Y}{\rho}} \tag{14.3}$$

The velocity of sound depends on the temperature of the medium. Equation 14.4 shows the temperature dependence in air where T is the absolute (Kelvin) temperature and 331 m/s is the speed of sound in air at 0 °C.

$$v = (331 \text{ m / s})\sqrt{\frac{T}{273 \text{ K}}} \tag{14.4}$$

The intensity of a wave is the rate at which energy flows across a unit area, A, in a plane perpendicular to the direction of travel of the wave. The SI units of intensity, I, are watts per square meter, W / m^2.

$$I \equiv \frac{\text{power}}{\text{area}} = \frac{\mathcal{P}}{A} \tag{14.6}$$

At a frequency of 1000 Hz, the faintest sound detectable by the human ear (**threshold of hearing**) has an intensity of about 10^{-12} W / m^2. An intensity of 1 W / m^2 is considered to be the greatest intensity which the ear can tolerate (**the threshold of pain**).

Comment on sound intensity.

The decibel scale is a logarithmic intensity scale. On this scale, the unit of sound intensity is the decibel, dB. The constant I_0 **is a reference intensity**, chosen to coincide with the threshold of hearing.

$$\beta \equiv 10 \log\left(\frac{I}{I_0}\right) \qquad (14.7)$$

$$I_0 = 1.0 \times 10^{-12} \text{ W / m}^2$$

In order to determine the decibel level corresponding to two different sources sounded simultaneously, first find the individual intensities I_1 and I_2 in W / m^2 and add these values to obtain the combined intensity $I = I_1 + I_2$. Finally, use Equation 14.7 to convert the intensity I to the decibel scale.

Comment on the decibel scale.

The intensity of a **spherical wave produced by a point source** is inversely proportional to the square of the distance from the source.

$$I = \frac{\text{average power}}{\text{area}} = \frac{\mathcal{P}_{av}}{4\pi r^2} \qquad (14.8)$$

The apparent change in frequency heard by an observer whenever there is relative motion between the source and the observer is called the Doppler effect. When using Equation 14.11 the values for v_o and v_s are each substituted with either a positive or negative sign depending on the direction of their velocity. If the observer or the source moves toward the other, that velocity is substituted with a positive sign. If observer or source moves away from the other, the corresponding velocity is substituted into Equation 14.11 with a negative sign. Also, it is important to remember that v_o (velocity of the observer) and v_s (velocity of the source) are **each measured relative to the medium in which the sound travels.**

$$f' = f\left(\frac{v + v_o}{v - v_s}\right) \qquad (14.11)$$

f = frequency of the source
f' = observed frequency
v = velocity of sound
v_o = velocity of observer
v_s = velocity of source

A series of standing wave patterns called normal modes can be excited in a **stretched string (fixed at both ends)**. Each mode corresponds to a characteristic frequency and wavelength.

$$\lambda_n = \frac{2L}{n}$$

$$f_n = nf_1 = \frac{n}{2L}\sqrt{\frac{F}{\mu}} \qquad (14.18)$$

where $n = 1, 2, 3, \ldots$

The frequency f_1 corresponding to $n = 1$, is the **fundamental frequency**, or first harmonic, and is the lowest frequency for which a standing wave is possible. The higher frequencies ($n = 2, 3, 4, \ldots$) form a harmonic series and are integral multiples of the fundamental frequency.

Comment on standing waves in strings.

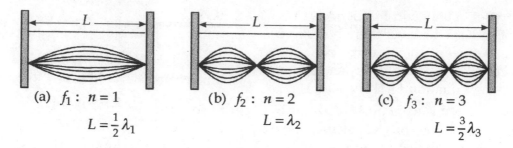

(a) f_1: $n = 1$
$$L = \frac{1}{2}\lambda_1$$

(b) f_2: $n = 2$
$$L = \lambda_2$$

(c) f_3: $n = 3$
$$L = \frac{3}{2}\lambda_3$$

Standing waves in a stretched string of length L fixed at both ends. The normal frequencies of vibration form a harmonic series: (a) the fundamental frequency, or first harmonic, (b) the second harmonic, and (c) the third harmonic.

In a pipe open at both ends, the natural frequencies of vibration form a series in which all harmonics (integer multiples of the fundamental) are present. Note that this series of frequencies is the same as the series of frequencies produced in a string fixed at both ends.

$$f_n = n\left(\frac{v}{2L}\right) \qquad (14.19)$$

where $n = 1, 2, 3, \ldots$

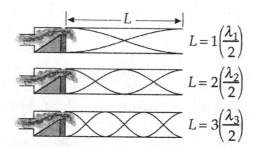

$$L = 1\left(\frac{\lambda_1}{2}\right)$$

$$L = 2\left(\frac{\lambda_2}{2}\right)$$

$$L = 3\left(\frac{\lambda_3}{2}\right)$$

In a pipe open at one end, only the odd harmonics (odd multiples of the fundamental) are possible.

$$f_n = n\left(\frac{v}{4L}\right) \qquad (14.20)$$

where $n = 1, 3, 5, \ldots$

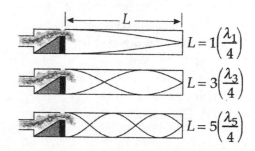

$$L = 1\left(\frac{\lambda_1}{4}\right)$$

$$L = 3\left(\frac{\lambda_3}{4}\right)$$

$$L = 5\left(\frac{\lambda_5}{4}\right)$$

SUGGESTIONS, SKILLS, AND STRATEGIES

When making calculations using Equation 14.7 which defines the intensity of a sound wave on the decibel scale, the properties of logarithms must be kept clearly in mind.

In order to determine the decibel level corresponding to two sources sounded simultaneously, you must first find the intensity, I, of each source in W / m^2, add these values, and then convert the resulting intensity to the decibel scale. As an illustration of this technique, note that if two sounds of intensity 40 dB and 45 dB are sounded together, the intensity level of the combined sources **is 46.2 dB (not 85 dB)**.

The most likely error in using Equation 14.11 to calculate the Doppler frequency shift due to relative motion between a sound source and an observer is due to using the incorrect algebraic sign for the velocity of either the observer or the source. Remember the following relationship between the relative velocity of the source and observer, and the corresponding Doppler frequency shift: the word **toward** is associated with an **increase** in frequency and the words **away from** are associated with a **decrease** in frequency.

REVIEW CHECKLIST

▷ Describe the harmonic displacement and pressure variation as functions of time and position for a harmonic sound wave.

▷ Calculate the speed of sound in various media in terms of appropriate elastic properties (these can include bulk modulus, Young's modulus, and the pressure-volume relationships of an ideal gas) and the corresponding inertial properties (usually the mass density).

▷ Understand the basis of the logarithmic intensity scale (decibel scale) and convert intensity values (given in W / m^2) to loudness levels on the dB scale. Determine the intensity ratio for two sound sources whose decibel levels are known. Calculate the intensity of a point source wave at a given distance from the source.

▷ Describe the various situations under which a Doppler shifted frequency is produced. Calculate the apparent frequency for a given actual frequency for each of the various possible relative motions between source and observer.

▷ Describe in both qualitative and quantitative terms the conditions which produce standing waves in a stretched string and in an open or closed air column pipe.

SOLUTIONS TO SELECTED END-OF-CHAPTER PROBLEMS

7.   You are watching a pier being constructed on the far shore of a saltwater inlet when some blasting occurs. You hear the sound in the water 4.50 s before it reaches you through the air. How wide is the inlet? (**Hint:** See Table 14.1. Assume the air temperature is 20 °C.)

Solution

From Table 14.1, the speed of sound in the saltwater is $v_w = 1530$ m / s.

At $$T = 20 \text{ °C} = 293 \text{ K}$$

the speed of sound in air is $v_a = (331 \text{ m / s})\sqrt{\dfrac{T}{273 \text{ K}}} = (331 \text{ m / s})\sqrt{\dfrac{293 \text{ K}}{273 \text{ K}}} = 343 \text{ m / s}$

If d is the width of the inlet, the time required for sound traveling in air to cross the inlet is $t_a = d / v_a$. Likewise, the required transit time for sound traveling through seawater is $t_w = d / v_w$.

Since it is given that $$t_a = t_w + 4.50 \text{ s}$$

we have $$\frac{d}{v_a} = \frac{d}{v_w} + 4.50 \text{ s}$$

or $$d = (4.50 \text{ s})\left(\frac{v_w v_a}{v_w - v_a}\right)$$

Therefore, the width of the inlet must be

$$d = (4.50 \text{ s})\left[\frac{(1530 \text{ m / s})(343 \text{ m / s})}{(1530 - 343) \text{ m / s}}\right]$$

$$d = 1.99 \times 10^3 \text{ m} = 1.99 \text{ km} \qquad\qquad \Diamond$$

13. A noisy machine in a factory produces sound with a level of 80 dB. How many identical machines could you add to the factory without exceeding the 90-dB limit?

Solution

The decibel level of a sound is given by

$$\beta = 10\log\left(\frac{I}{I_0}\right)$$

where I is the intensity of the sound wave and $I_0 = 1.0 \times 10^{-12}$ W / m^2 is a reference intensity.

Solving for the intensity gives $\qquad I = I_0 10^{\beta/10}$

The intensity of sound produced by one machine ($\beta = 80$ dB) is

$$I_1 = I_0 10^{80/10} = I_0 10^{8.0}$$

The intensity of sound necessary to reach a 90-dB level is

$$I = I_0 10^{90/10} = I_0 10^{9.0}$$

The number of machines which, acting together, would produce a 90-dB sound level is

$$N = \frac{I}{I_1} = \frac{I_0 10^{9.0}}{I_0 10^{8.0}} = 10$$

Since the factory already contains one machine, you can add 9 additional machines without exceeding the limit. $\qquad \Diamond$

---

17. A train sounds its horn as it approaches an intersection. The horn can just be heard at a level of 50 dB by an observer 10 km away. (a) What is the average power generated by the horn? (b) What intensity level of the horn's sound is observed by someone waiting at an intersection 50 m from the train? Treat the horn as a point source and neglect any absorption of sound by the air.

Solution

(a) If the train horn acts as a point source, the wave fronts are spherical with a surface area $A = 4\pi r^2$ at distance r from the horn. The average power emitted by a source can be expressed as $\mathcal{P} = IA$ where I is the intensity of the wave and A is the surface area of the wave front, both at the same distance r from the source.

The decibel level of a sound is defined as $\quad \beta \equiv 10\log(I/I_0)$

where I is the intensity of the sound and $\quad I_0 = 1.0 \times 10^{-12}$ W / m^2

Solving this equation for the intensity gives $\quad I = I_0\left(10^{\beta/10}\right)$

At $r = 10$ km $= 1.0 \times 10^4$ m from the train horn, the sound level is 50 dB.

Therefore, the intensity of the sound is

$$I = \left(1.0 \times 10^{-12} \text{ W / m}^2\right)10^{50/10} = \left(1.0 \times 10^{-12} \text{ W / m}^2\right)10^{5.0} = 1.0 \times 10^{-7} \text{ W / m}^2$$

and the average power generated by the horn must be

$$\mathcal{P} = IA = I\left(4\pi r^2\right) = \left(1.0 \times 10^{-7} \text{ W / m}^2\right)\left[4\pi\left(1.0 \times 10^4 \text{ m}\right)^2\right] = 1.3 \times 10^2 \text{ W} \qquad \Diamond$$

(b) At a distance of $r = 50$ m, the intensity of the sound is

$$I = \frac{\mathcal{P}}{A} = \frac{\mathcal{P}}{4\pi r^2} = \frac{1.26 \times 10^2 \text{ W}}{4\pi(50 \text{ m})^2} = 4.0 \times 10^{-3} \text{ W / m}^2$$

and the decibel level is

$$\beta = 10\log\left(\frac{I}{I_0}\right) = 10\log\left(\frac{4.0 \times 10^{-3} \text{ W / m}^2}{1.0 \times 10^{-12} \text{ W / m}^2}\right) = 10\log\left(4.0 \times 10^9\right) = 96 \text{ dB} \qquad \Diamond$$

25. An alert physics student stands beside the tracks as a train rolls slowly past. He notes that the frequency of the train whistle is 442 Hz when the train is approaching him and 441 Hz when the train is receding from him. From this he can find the speed of the train. What value does he find?

Solution   According to the Doppler effect, a sound with a source frequency f will have an observed frequency of

$$f' = f\left(\frac{v + v_0}{v - v_s}\right)$$

where v is the speed of the sound wave, while v_0 and v_s are the speeds of the observer and the source respectively.

v_0 is considered positive when the observer moves **toward** the source and negative when the observer moves **away from** the source. Likewise, v_s is considered positive when the source moves **toward** the observer and negative when it moves **away from** the observer.

Since the student is at rest beside the tracks, $v_0 = 0$. As stated at the start of the Chapter 14 problem section in the textbook, we use $v = 345$ m / s as the speed of sound in air. When the train is approaching the student, $v_s > 0$. Thus, $v_s = +|v_s|$ and the Doppler effect equation gives

$$442\text{ Hz} = f\left(\frac{345\text{ m / s}}{345\text{ m / s} - |v_s|}\right) \qquad [1]$$

When the train is receding, $v_s < 0$ so $v_s = -|v_s|$

and the relation becomes $$441\text{ Hz} = f\left(\frac{345\text{ m / s}}{345\text{ m / s} + |v_s|}\right) \qquad [2]$$

Dividing Equation [1] by Equation [2] gives $\dfrac{442}{441} = \dfrac{345\text{ m / s} + |v_s|}{345\text{ m / s} - |v_s|}$

or $442(345\text{ m / s}) - 442|v_s| = 441(345\text{ m / s}) + 441|v_s|$

Solving for the train speed, $883|v_s| = 345$ m / s

and $|v_s| = 0.391$ m / s   ◊

229

33. A pair of speakers separated by 0.700 m are driven by the same oscillator at a frequency of 690 Hz. An observer, originally positioned at one of the speakers, begins to walk along a line perpendicular to the line joining the two speakers. (a) How far must the observer walk before reaching a relative maximum in intensity? (b) How far will the observer be from the speaker when the first relative minimum is detected in the intensity?

Solution

The wavelength of the sound emitted by the speakers is

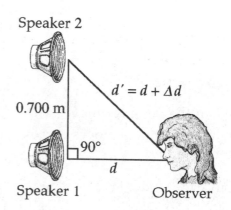

$$\lambda = \frac{v}{f} = \frac{345 \text{ m/s}}{690 \text{ Hz}} = 0.500 \text{ m}$$

When the observer is distance d from the first speaker, he is distance $d' = d + \Delta d$ from the second as shown in the sketch. If a relative maximum is to occur, the difference in distances the two sound waves have traveled from the sources to the observer, $\Delta d = d' - d$, must be a whole number of wavelengths.

For a relative minimum, this path difference must be an odd number of half wavelengths. Note that when the observer is at Speaker 1 (i.e., when $d = 0$), the path difference is $\Delta d = 0.700 \text{ m} = 1.40\lambda$. As the observer moves away from this speaker, Δd decreases and approaches zero as d approaches infinity. Along this route, Δd ranges from 1.40λ to 0, so the observer will experience only one relative maximum (at $\Delta d = \lambda$) and one relative minimum (at $\Delta d = \lambda/2$).

(a)   When the relative maximum is reached, $\Delta d = \lambda = 0.500 \text{ m}$

Applying the Pythagorean theorem to the right triangle shown in the sketch,

$$(d + 0.500 \text{ m})^2 = (0.700 \text{ m})^2 + d^2$$

Expanding and simplifying this gives $(1.00 \text{ m})d + 0.250 \text{ m}^2 = 0.490 \text{ m}^2$

or $d = \dfrac{0.490 \text{ m}^2 - 0.250 \text{ m}^2}{1.00 \text{ m}}$

Thus, the relative maximum is located at $d = 0.240 \text{ m} = 24.0 \text{ cm}$

in front of Speaker 1   ◊

230

(b)   At the relative minimum, $\Delta d = \lambda/2 = 0.250$ m

and the Pythagorean theorem gives $(d + 0.250 \text{ m})^2 = (0.700 \text{ m})^2 + d^2$

Thus, the location of the relative minimum is

$$d = \frac{0.490 \text{ m}^2 - 0.0625 \text{ m}^2}{0.500 \text{ m}} = 0.855 \text{ m} = 85.5 \text{ cm} \quad \text{from Speaker 1.} \qquad \Diamond$$

---

37.   Two speakers are driven by a common oscillator at 800 Hz and face each other at a distance of 1.25 m. Locate the points along a line joining the two speakers where relative minima of pressure amplitude would be expected. (Use $v = 343$ m / s.)

Solution   Since the speakers are driven by a common oscillator, they must vibrate in phase with each other. Thus, the point halfway between them (being equidistant from the two sources) must be an anti-node in any standing wave pattern formed.

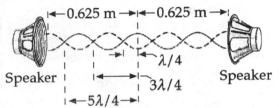

Relative minima (i.e., nodes) are located on either side of this central anti-node at a distance of

$$\frac{\lambda}{4} = \frac{1}{4}\left(\frac{v}{f}\right) = \frac{343 \text{ m / s}}{4(800 \text{ Hz})} = 0.107 \text{ m} \quad \text{from the midpoint.}$$

This means that they are located at distances of

$$d = 0.625 \text{ m} \pm 0.107 \text{ m} = 5.18 \text{ m} \quad \text{and } 0.732 \text{ m} \quad \text{from either speaker} \qquad \Diamond$$

A second pair of nodes will be found at a distance of $3\lambda/4 = 3(0.107 \text{ m})$ on either side of the midpoint. The positions of these nodes are then located at distances of

$$d = 0.625 \text{ m} \pm 3(0.107 \text{ m}) = 0.303 \text{ m} \quad \text{and } 0.947 \text{ m from either speaker} \qquad \Diamond$$

Finally a pair of nodes will be found at a distance of $5\lambda/4 = 5(0.107 \text{ m})$ on either side of the midpoint. These points are located at distances of

$$d = 0.625 \text{ m} \pm 5(0.107 \text{ m}) = 0.089 \text{ m} \quad \text{and } 1.16 \text{ m from either speaker} \qquad \Diamond$$

---

40. In the arrangement shown in Figure P14.40, an object of mass $m = 5.0$ kg, hangs from a cord around a light pulley. The length of the cord between point P and the pulley is $L = 2.0$ m. (a) When the vibrator is set to a frequency of 150 Hz, a standing wave with six loops is formed. What must be the linear mass density of the cord? (b) How many loops (if any) will result if m is changed to 45 kg? (c) How many loops (if any) will result if m is changed to 10 kg?

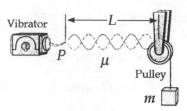

Figure P14.40

Solution

(a) Each loop in the cord is one-half wavelength long. Thus, when a standing wave pattern with six loops forms in the cord, $L = 6(\lambda / 2)$,

and the wavelength is $\qquad \lambda = L / 3 = (2.0 \text{ m}) / 3 = 0.67 \text{ m}$

The speed of the waves in the cord is $\quad v = \lambda f = (0.67 \text{ m})(150 \text{ Hz}) = 100 \text{ m/s}$

This speed is also given by $\qquad v = \sqrt{F / \mu}$

where F is the tension in the cord and μ is the linear mass density. Since the 5.0-kg object is in equilibrium, the cord's tension must equal the weight of this object,

or $\qquad\qquad\qquad\qquad\qquad F = mg = 49 \text{ N}$

Therefore, the mass per unit length of the cord is

$$\mu = \frac{F}{v^2} = \frac{49 \text{ N}}{(100 \text{ m / s})^2} = 4.9 \times 10^{-3} \text{ kg / m} \quad \Diamond$$

(b) If the second object has mass $\qquad m = 45 \text{ kg}$

the tension in the cord will be $\qquad F = mg = (45 \text{ kg})(9.8 \text{ m/s}^2) = 4.4 \times 10^2 \text{ N}$

and the speed of the waves is $\qquad v = \sqrt{\frac{F}{\mu}} = \sqrt{\frac{4.4 \times 10^2 \text{ N}}{4.9 \times 10^{-3} \text{ kg / m}}} = 3.0 \times 10^2 \text{ m / s}$

The wavelength is then $\qquad \lambda = \frac{v}{f} = \frac{3.0 \times 10^2 \text{ m / s}}{150 \text{ Hz}} = 2.0 \text{ m}$

The number of loops that now fit in the length of the cord is

$$n = \frac{L}{\lambda/2} = \frac{2.0 \text{ m}}{1.0 \text{ m}} = 2$$

This is a whole number, so the cord forms a standing wave of 2 loops. $\qquad \Diamond$

(c)   If $m = 10$ kg , then

$$F = mg = 98 \text{ N}$$

and the speed is

$$v = \sqrt{\frac{98 \text{ N}}{4.9 \times 10^{-3} \text{ kg/m}}} = 1.4 \times 10^2 \text{ m/s}$$

Therefore, the wavelength is

$$\lambda = \frac{v}{f} = \frac{1.4 \times 10^2 \text{ m/s}}{150 \text{ Hz}} = 0.94 \text{ m}$$

The number of half-wavelengths in the length of the cord is

$$n = \frac{L}{\lambda/2} = \frac{2.0 \text{ m}}{(0.94 \text{ m})/2} = 4.2$$

Since this is **not an integer**, resonance does not occur and no standing wave pattern is produced.   ◊

---

47.   A pipe open at both ends has a fundamental frequency of 300 Hz when the temperature is 0 °C. (a) What is the length of the pipe? (b) What is the fundamental frequency at a temperature of 30 °C?

Solution   A standing wave pattern in a pipe open at both ends must have an anti-node at each end.   In the fundamental mode of vibration (i.e., for the lowest frequency and hence longest wavelength capable of producing resonance), the standing wave pattern is as shown in the sketch.

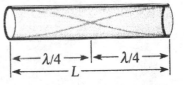

Therefore, the length of the pipe is

$$L = \lambda / 4 + \lambda / 4 = \lambda / 2$$

(a)   The speed of sound in air at absolute temperature T is

$$v = (331 \text{ m/s})\sqrt{T/273 \text{ K}}$$

Thus, at $T = 0 \text{ °C} = 273 \text{ K}$,

$$v = 331 \text{ m/s}$$

If the fundamental frequency of the pipe is 300 Hz,

the length of the pipe is

$$L = \frac{\lambda}{2} = \frac{v}{2f} = \frac{331 \text{ m/s}}{2(300 \text{ Hz})} = 0.552 \text{ m}$$ ◊

(b)   At

$$T = 30 \text{ °C} = 303 \text{ K}$$

the speed of sound is

$$v = (331 \text{ m/s})\sqrt{\frac{303 \text{ K}}{273 \text{ K}}} = 349 \text{ m/s}$$

The change in the length of the pipe as the temperature rises from 0 °C to 30 °C is negligible in comparison to the total length. Thus, the length of the pipe is still $L = 0.552$ m and, in the fundamental resonance mode, $\lambda = 2L$. The fundamental frequency at this temperature is then

$$f = \frac{v}{\lambda} = \frac{349 \text{ m / s}}{2(0.552 \text{ m})} = 316 \text{ Hz} \qquad \Diamond$$

51. Two train whistles have identical frequencies of 180 Hz. When one train is at rest in the station, sounding its whistle, a beat frequency of 2 Hz is heard from a moving train. What two possible speeds and directions can the moving train have?

Solution The beat frequency is equal to the difference in the detected frequencies of the two sounds. Thus, the frequency the observer detects for the sound from the whistle on the moving train must be either

$$f' = 180 \text{ Hz} - 2 \text{ Hz} = 178 \text{ Hz}$$

or

$$f' = 180 \text{ Hz} + 2 \text{ Hz} = 182 \text{ Hz}$$

The Doppler effect is responsible for the difference in the detected frequencies.

The frequency a stationary observer detects from a moving sound source is

$$f' = f\left(\frac{v}{v - v_s}\right)$$

where v is the speed of sound, and f is the frequency heard when the source is stationary relative to the observer.

The speed of the source is positive $(v_s = |v_s|)$ when the source is approaching the observer and negative $(v_s = -|v_s|)$ when the source is moving away from the observer.

Note that if the source is approaching the observer, $f' > f$, and $f' < f$ if the source is receding. In this case, $f = 180$ Hz and the speed of sound is assumed to be $v = 345$ m/s.

234

The two possible situations are as follows:

(1) If the moving train is approaching the station, the Doppler effect equation becomes

$$182 \text{ Hz} = (180 \text{ Hz})\left(\frac{345 \text{ m / s}}{345 \text{ m / s} - |v_s|}\right)$$

or

$$345 \text{ m / s} - |v_s| = \left(\frac{180}{182}\right)(345 \text{ m / s})$$

Thus, the train moves toward the station with speed

$$|v_s| = \left(1 - \frac{180}{182}\right)(345 \text{ m / s}) = 3.79 \text{ m / s} \qquad \Diamond$$

(2) If the train is moving away from the station,

the equation yields

$$178 \text{ Hz} = (180 \text{ Hz})\left(\frac{345 \text{ m / s}}{345 \text{ m / s} + |v_s|}\right)$$

or

$$345 \text{ m / s} + |v_s| = \left(\frac{180}{178}\right)(345 \text{ m / s})$$

In this case, the train recedes from the station with speed

$$|v_s| = \left(\frac{180}{178} - 1\right)(345 \text{ m / s}) = 3.88 \text{ m / s} \qquad \Diamond$$

---

54. If a human ear canal can be thought of as resembling an organ pipe, closed at one end, that resonates at a fundamental frequency of 3000 Hz, what is the length of the canal? (Use normal body temperature 37 °C for your determination of the speed of sound in the canal.)

Solution A standing wave pattern in a pipe that is open at one end but closed at the other, must have an anti-node at the open end and a node at the closed end of the pipe. In the fundamental resonance mode (longest wavelength and lowest frequency that can produce a standing wave) for such a pipe, the wavelength of the sound is $\lambda = 4L$ as illustrated in the sketch.

Normal body temperature is $T = 37 \,°C = 310 \text{ K}$, and the speed of sound in air at this temperature is

$$v = (331 \text{ m / s})\sqrt{\frac{T}{273 \text{ K}}} = (331 \text{ m / s})\sqrt{\frac{310 \text{ K}}{273 \text{ K}}} = 353 \text{ m / s}$$

The wavelength of a 3000-Hz sound wave at this temperature is

$$\lambda = \frac{v}{f} = \frac{353 \text{ m/s}}{3000 \text{ hz}} = 0.118 \text{ m} = 11.8 \text{ cm}$$

The length of the human ear canal that resonates at a fundamental frequency of 3000 Hz is therefore

$$L = \frac{\lambda}{4} = \frac{11.8 \text{ cm}}{4} = 2.94 \text{ cm}$$ ◊

---

59. On a workday the average decibel level of a busy street is 70 dB, with 100 cars passing a given point every minute. If the number of cars is reduced to 25 every minute on a weekend, what is the decibel level of the street?

Solution

The decibel level of a sound is given by $\beta = 10\log(I/I_0)$ where I is the   intensity of the sound and I_0 is a reference intensity.

On the weekend, there are one-fourth as many cars passing per minute as on a week day. Thus, the expected sound intensity, I_2, on the weekend should be one-fourth the sound intensity, I_1, on a week day. The difference in the decibel levels on a week day and on the weekend will be

$$\beta_1 - \beta_2 = 10\log\left(\frac{I_1}{I_0}\right) - 10\log\left(\frac{I_2}{I_0}\right) = 10\log\left(\frac{I_1}{I_2}\right) = 10\log(4) = 6 \text{ dB}$$

The decibel level on the weekend is $\beta_2 = \beta_1 - 6 \text{ dB} = 70 \text{ dB} - 6 \text{ dB} = 64 \text{ dB}$ ◊

---

67. By proper excitation, it is possible to produce both longitudinal and transverse waves in a long metal rod. In a particular case, the rod is 150 cm long and 0.200 cm in radius and has a mass of 50.9 g. Young's modulus for the material is 6.80×10^{10} Pa. Determine the required tension in the rod so that the ratio of the speed of longitudinal waves to the speed of transverse waves is 8.

Solution

The speed of longitudinal waves in a solid is given by $v_L = \sqrt{Y/\rho}$ where Y is the Young's modulus and ρ is the density (mass per unit volume) of the solid.

Transverse waves travel along a long, thin medium such as a rod with a speed $v_t = \sqrt{F/\mu}$ where F is the tension in this medium and μ is the mass per unit length.

If $v_L = 8v_t$, then

$$\sqrt{Y/\rho} = 8\sqrt{F/\mu}$$

or the required tension in the rod is

$$F = (Y/64)(\mu/\rho)$$

The volume of a rod is $V = AL$ where A is the cross-sectional area of the rod and L is its length. Therefore, for a uniform cylindrical rod, the ratio of densities that appears in the equation for the tension is

$$\frac{\mu}{\rho} = \frac{(m/L)}{(m/V)} = \frac{V}{L} = \frac{AL}{L} = A = \pi r^2$$

Thus, the needed tension is

$$F = \left(\pi r^2\right)Y/64$$

This rod has a radius of

$$r = 2.00 \times 10^{-1} \text{ cm} = 2.00 \times 10^{-3} \text{ m}$$

and is made from a material with a Young's modulus of

$$Y = 6.80 \times 10^{10} \text{ Pa} = 6.80 \times 10^{10} \text{ N}/\text{m}^2$$

The required tension in the rod is then

$$F = \frac{\pi\left(2.00 \times 10^{-3} \text{ m}\right)^2 \left(6.80 \times 10^{10} \text{ N/m}^2\right)}{64} = 1.34 \times 10^4 \text{ N} \qquad \diamond$$